AF823732

Gynecologic and Obstetric Investigation

Founded 1895 as 'Monatsschrift für Geburtshilfe und Gynäkologie',
continued 1946–1969 as 'Gynaecologia' and 1970–1977 as 'Gynecologic Investigation'

KARGER

Printed in Switzerland on acid-free paper by Reinhardt Druck, Basel

Appears 6-weekly: 2 volumes per year (8 issues)

Guidelines for Authors

Gynecologic and
Obstetric Investigation

Submission

Only original papers, carefully selected case reports, and invited reviews written in English are considered and should be sent to:

Prof. G. Zador
S. Karger AG
Editorial Office 'Gynecologic and
Obstetric Investigation'
P.O. Box
CH–4009 Basel (Switzerland)
Telephone +41 61 306 1356
Fax +41 61 306 1434
E-mail a.scholtz@karger.ch

Mailing address for courier deliveries only:
Allschwilerstr. 10
CH–4055 Basel (Switzerland)

Manuscripts should be submitted in triplicate (with three sets of illustrations of which one is an original), typewritten double-spaced on one side of the paper, with a wide margin.

Conditions

All manuscripts are subject to editorial review. *Original Papers* should not exceed 4 printed pages (12 manuscript pages), including tables, illustrations and references. *Articles on current investigation* are limited to 2 printed pages (6 manuscript pages). *Case Reports* are carefully selected and limited to 2 printed pages (6 manuscript pages): this includes statements of the problem, patient history, diagnosis, treatment, results and any other information pertinent to the case and maximum 5 references. *Reviews* are usually commissioned, however, the Editor-in-Chief is happy to discuss potential articles with authors who would like to contribute. Submission of an article for publication implies the transfer of the copyright from the author to the publisher upon acceptance. Accepted papers become the permanent property of 'Gynecologic and Obstetric Investigation' and may not be reproduced by any means, in whole or in part, without the written consent of the publisher. It is the author's responsibility to obtain permission to reproduce illustrations, tables, etc. from other publications.

All manuscripts must be accompanied by a covering letter signed by all authors. Assurance should be given that the manuscript is not under simultaneous consideration by any other publication.

All manuscripts originating from non-English speaking countries must be revised by a professional linguistic reviewer and it must be evident from the covering letter that this has been done.

Good clinical practice: It has become mandatory that every trial in humans must first obtain approval from an independent Ethics Committee and formal, informed consent from the patients before they participate in a clinical study or experiment. In order to avoid unnecessary delay with the review of manuscripts, authors are kindly asked to state, preferably in the Materials and Methods section, that approval and informed consent have been obtained. These two statements must also appear on the covering letter which accompanies every manuscript and is signed by each author.

Arrangement

All pages should be consecutively numbered beginning with the title page, then the text, acknowledgements, references and legends to figures. The text in original papers should be divided under the headings: Abstract, Introduction, Material(s) and Method(s), Results, and Discussion. The text in case reports should be divided under the headings: Introduction, Case Report(s) and Conclusion(s).

Title page: The first page of each paper should indicate the title (main title underlined), the authors' names, and the institute where the work was conducted. A short title for use as running head as well as the full address of the author to whom correspondence should be sent are also required.

Full address: The exact postal address complete with postal code must be given at the bottom of the title page. Please also supply phone and fax numbers, as well as your e-mail address.

Key words: For indexing purposes, a list of 3–10 key words in English is essential.

Abstract: The first page of the text should include an abstract of up to 10 lines.

Small type: Paragraphs which can or must be set in smaller type (case histories, test methods, etc.) should be indicated with a 'p' (petit) in the margin on the left-hand side.

Footnotes: Avoid footnotes. When essential, they are numbered consecutively and typed at the foot of the appropriate page.

Tables and illustrations: Tables and illustrations (both numbered in Arabic numerals) should be prepared on separate sheets. Tables require a heading and figures a legend, also prepared on a separate sheet. For the reproduction of illustrations, only good drawings and original photographs can be accepted; negatives or photocopies cannot be used. Due to technical reasons, figures with a screen background should not be submitted. When possible, group several illustrations on one block for reproduction (max. size 181 × 223 mm) or provide crop marks. On the back of each illustration, indicate its number, the author's name, and 'top' with a soft pencil.

Color illustrations: Up to 6 color illustrations per page can be integrated with the text at the reduced price of CHF 660.– / USD 545.00 per page. Color illustrations are reproduced at the author's expense.

References: In the text identify references by Arabic numerals [in square brackets]. Material submitted for publication but not yet accepted should be noted as 'unpublished data' and not be included in the reference list. The list of references should include only those publications which are cited in the text. Do not alphabetize; number references in the order in which they are first mentioned in the text. The surnames of the authors followed by initials should be given. There should be no punctuation other than a comma to separate the authors. Cite all authors, 'et al' is not sufficient. Abbreviate journal names according to the Index Medicus system. (Also see International Committee of Medical Journal Editors: Uniform requirements for manuscripts submitted to biomedical journals. N Engl J Med 1997;336:309–315.)

Examples

(a) Papers published in periodicals: Sun J, Koto H, Chung KF: Interaction of ozone and allergen challenges on bronchial responsiveness and inflammation in sensitised guinea pigs. Int Arch Allergy Immunol 1997;112:191–195.

(b) Monographs: Matthews DE, Farewell VT: Using and Understanding Medical Statistics, ed 3, revised. Basel, Karger, 1996.

(c) Edited books: Parren PWHI, Burton DR: Antibodies against HIV-1 from phage display libraries: Mapping of an immune response and progress towards antiviral immunotherapy; in Capra JD (ed): Antibody Engineering. Chem Immunol. Basel, Karger, 1997, vol 65, pp 18–56.

Page charges

There is no page charge for papers of 4 or fewer printed pages (including tables, illustrations and references). Each additional complete or partial page is charged to the author at CHF 275.– / USD 196.00. The allotted size of a paper is equal to approx. 12 manuscript pages (including tables, illustrations and references).

Galley proofs

Unless indicated otherwise, galley proofs are sent to the first-named author and should be returned with the least possible delay. Alterations made in galley proofs, other than the correction of printer's errors, are charged to the author. No page proofs are supplied.

Reprints

Order forms and a price list are sent with the galley proofs. Orders submitted after the issue is printed are subject to considerably higher prices.

KARGER

Fax + 41 61 306 12 34
E-Mail karger@karger.ch
www.karger.com

The Guidelines for Authors are available at:
www.karger.com/journals/goi/goi_gl.htm

General Information

Gynecologic and Obstetric Investigation

Publication data: 'Gynecologic and Obstetric Investigation' is published 8 times a year. Volumes 47–48 with 4 issues each appear in 1999.
Subscription rates: Subscriptions run for a full calendar year. Prices are given per volume.
Personal subscription:
Subscription + postage and handling
CHF 285.– + CHF 15.70 Europe
CHF 285.– + CHF 23.70 Overseas
USD 218.00 + USD 15.70
(Must be in the name of, billed to, and paid by an individual. Order must be marked 'personal subscription'.)
Institutional subscription:
Subscription + postage and handling
CHF 570.– + CHF 19.60 Europe
CHF 570.– + CHF 29.60 Overseas
USD 436.00 + USD 19.60
Airmail postage: CHF 24.–, USD 20.00.
(Extra per volume.)
Discount subscription prices:
Society for Gynecologic Investigation
Microforms: Please direct all enquiries and orders to: University Microfilms Inc.
300 North Zeeb Road, P.O. Box 1346
Ann Arbor, MI 48106-1346 (USA)

Subscription orders: Orders can be placed at agencies, bookstores, directly with the Publisher
S. Karger AG, P.O. Box,
CH–4009 Basel (Switzerland)
or with any of the following distributors:
Deutschland: S. Karger GmbH, Postfach
D–79095 Freiburg
France: Librairie Luginbühl
36, bd de Latour-Maubourg, F–75007 Paris
Great Britain, Ireland: S. Karger Publishers
58 Grove Hill Road
Tunbridge Wells, Kent TN1 1SP
USA: S. Karger Publishers, Inc.
26 West Avon Road, P.O. Box 529
Farmington, CT 06085
India, Bangladesh, Sri Lanka:
S. Karger India Subscription Agency
B–5/132 Safdarjung Enclave
New Delhi-110029
Japan: Katakura Libri, Inc.
2F, Marushima Bldg, 4-6-17, Yushima,
Bunkyo-ku, Tokyo 113-0034
Change of address: Both old and new address should be sent to the subscription source.
Single issues and back volumes: Information can be obtained through the Publisher.
Advertising: Correspondence and rate requests should be addressed to the Publisher.

Bibliographic indices: This journal is regularly listed in bibliographic services, including *Current Contents®*.
Book Reviews: Books are accepted for review by special agreement.
Copying: This journal has been registered with the Copyright Clearance Center (CCC), as indicated by the code appearing on the first page of each article.
For readers in the US, this code signals consent for copying of articles for personal or internal use, or for the personal or internal use of specific clients, provided that the stated fee is paid per copy directly to
Copyright Clearance Center Inc.
222 Rosewood Drive
Danvers, MA 01923 (USA)
A copy of the first page of the article must accompany payment. Consent does not extend to copying for general distribution, for promotion, for creating new works, or for resale. In these cases, specific written permission must be obtained from the copyright owner,
S. Karger AG,
P.O. Box, CH–4009 Basel (Switzerland).

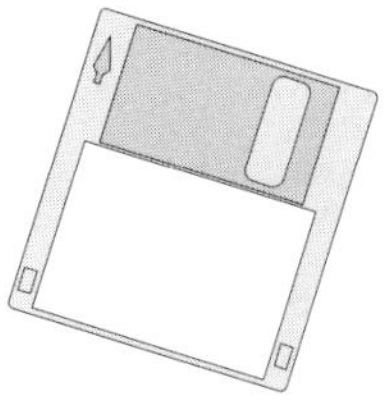

Disk Submission

A **comprehensive version** of these guidelines for the **submission of manuscripts and illustrations on disk** can be accessed online under **www.karger.com/journals/disksub.htm** or requested as a leaflet from the editorial department.

Please follow the general rules on style and arrangement in the **"Guidelines for Authors"** of the journal. Submit a **double-spaced printout** of your manuscript matching the file on disk exactly.

Technical Information

Store only data pertinent to this manuscript. Include your fonts (screen and printer fonts for Macintosh). For a **text only** manuscript, use a 3.5 inch disk. For **text and illustrations,** use 3.5 or 5.25 Magneto Optical Disc; CD-ROM (DOS or hybrid); iomage Zip/DOS (100 MB), or 44 MB, 88 MB or 200 MB SyQuest. Compress extensive data using ZIP format.

Data Description

Label your disk with the **title of the journal or book** for which it is submitted; a **short version** of the paper's title; your **name,** postal and e-mail address, fax and phone number; **hard and software** used, including the version; and the **complete list of the file names.** File names consist of eight characters before the dot identifying the file and three characters after the dot identifying the application.

Text and Format

The **preferred word-processing package is MS-Word.** Other commonly used PC text programs and ASCII formats are accepted.
Update your data when you revise subsequent to the reviewer's or editor's decision and mark revisions clearly. Manuscripts accepted for publication are subject to **copy editing:** changes and/or questions will be marked in the manuscript, which will be returned to you with your disk and the galley proofs for checking.
Enter your text continuously flush left. Do not use any layout functions. Do not split words at the end of a line. **Do not use indentations.** Use a blank line before and a hard return at the end of paragraphs. **Arrange headings flush left.** Insert only **one space after punctuation marks.** Insert Greek letters, mathematical symbols etc. with your word-processing program. Use **boldface and italics** as well as sub- and superscript where appropriate.
Tables are part of the text and should be placed at the end of the text file. Separate the individual parts of a table by lines. Use the tabulator instead of spaces. Format operation data on one line in the same column. **Legends to illustrations are part of the text** and should also be listed at the end of the text file.

Illustrations

Store illustrations as separate files: do not embed them in your text. Export **line drawings/vector graphics** as TIF, EPS or WMF format. Use Photoshop for processing and retouching **scanned halftone images.** Save the original scan and the processed version. Export **black and white or color images** as TIF or EPS format in their anticipated size in print.
For detailed information on how to submit illustration data on disk please consult the leaflet or **www.karger.com/journals/disksub.htm.**

KARGER
Fax +41 61 306 12 34
E-Mail karger@karger.ch
www.karger.com

The Journal Home Page is available at:
www.karger.com/journals/goi

KI 98 DISK

Expert guidelines
for prevention and diagnosis

Virus Infections in Pregnancy

Editor: Uwe G. Liebert

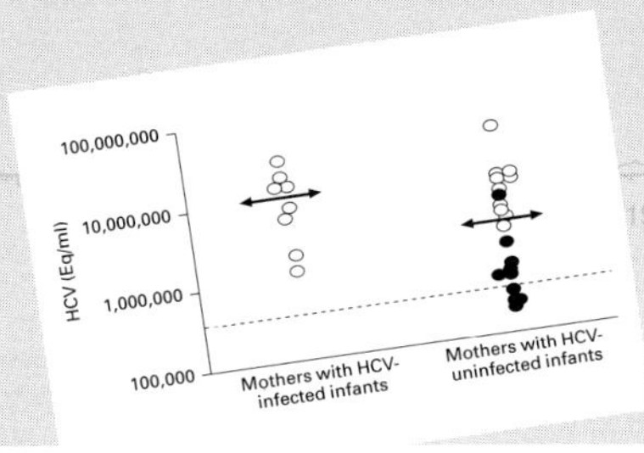

Virus infections in pregnancy do not only cause immediate morbidity in the mother, fetus and newborn, but can produce long-term damage or malformations in the unborn child. Vaccination can protect against vertical transmission of some viruses, but in most cases effective prevention of congenital infections and prognosis can only be achieved by rapid and accurate laboratory diagnosis.

In this issue of *Intervirology* international experts review the current state of techniques available for the prenatal diagnosis of intrauterine rubella virus, cytomegalovirus and parvovirus infections as well as diagnostic procedures for neonatal infections with herpes simplex virus, HIV, varicella-zoster virus, hepatitis B and C viruses, and human papillomavirus. While specific IgM may be helpful in most cases, virus detection by PCR or virus isolation are the methods of choice wherever possible. The timely and accurate diagnosis of viral infections requires a close collaboration between clinicians and virologists, which is particularly important for the collection of specimens from both fetus/infant and mother, and for the optimal configuration of diagnostic laboratory techniques.

The recommendations for diagnosis, risk assessment, prophylaxis and treatment of viral infections provided in this issue make it a valuable source of information for virologists, gynecologists, obstetricians, pediatricians and specialists in laboratory medicine.

Contents

Liebert, U.G.: Editorial: Viruses and Pregnancy
Lazzarotto, T. et al.: Cytomegalovirus Infection in Pregnancy: A Still Complicated Diagnostic Problem
Grangeot-Keros, L. et al.: Should We Routinely Screen for Cytomegalovirus Antibody during Pregnancy?
Katow, S.: Rubella Virus Genome Diagnosis during Pregnancy and Mechanism of Congenital Rubella
Pustowoit, B.; Liebert, U.G.: Predictive Value of Serological Tests in Rubella Virus Infection during Pregnancy
Eis-Hübinger, A.M. et al.: Parvovirus B19 Infection in Pregnancy
Desselberger, U.: Herpes simplex Virus Infection in Pregnancy: Diagnosis and Significance
Sauerbrei, A.: Varicella-Zoster Virus Infections in Pregnancy
Papaevangelou, G. et al.: Hepatitis B Maternal-Fetal Transmission in Southern Europe
Berger, A. et al.: Human Immunodeficiency Virus and Hepatitis B Virus Infection in Pregnancy: Diagnostic Potential of Viral Genome Detection
Zanetti, A.R.: A Prospective Study on Mother-to-Infant Transmission of Hepatitis C Virus
Cason J. et al.: Transmission of Cervical Cancer-Associated Human Papilloma Viruses from Mother to Child
Horn, L.-C.; Röse, I.: Placental and Fetal Pathology in Intrauterine Viral Infections
Bohmer, R.M.: Fetal Cells from Maternal Blood: Purpose, Biological Questions, Technical Challenges

Intervirology
Editor: Braun, R.W. (Stuttgart)
ISSN 0300–5526
Special issue: Vol. 41, No. 4–5 (1998)
Included in subscription

Fields of Interest: Virology; Infectious Diseases; Fetal Medicine, Gynecology, Obstetrics, Pediatrics, Laboratory, Teratology

Virus Infections in Pregnancy
Editor: Liebert, U.G. (Leipzig)
92 p., 28 fig., 32 tab., soft cover, 1999
CHF 39.– / DEM 47.– / USD 34.00
Prices subject to change
DEM price for Germany, USD price for USA only
ISBN 3–8055–6872–X

KI 99227

KARGER

Gynecologic and
Obstetric Investigation

The Second Japan Conference on Endometriosis

Biology of Endometrium versus Endometriosis

Proceedings of the Second Japan Conference on Endometriosis
Hakone Prince Hotel, Kanagawa, Japan, May 8, 1999

Guest Editors
H. Hoshiai, Osaka-Sayama
Y. Taketani, Tokyo
N. Terakawa, Yonago

20 figures, 2 in color and 3 tables, 1999

KARGER
Basel · Freiburg · Paris · London · New York ·
New Delhi · Bangkok · Singapore · Tokyo · Sydney

The current meeting and publication were made possible by an educational grant from Takeda Chemical Industries, Ltd. Editorial assistance was provided by BIOMEDIS International Ltd.

S. Karger
Medical and Scientific Publishers
Basel · Freiburg · Paris · London
New York · New Delhi · Bangkok
Singapore · Tokyo · Sydney

Drug Dosage
The authors and the publisher have exerted every effort to ensure that drug selection and dosage set forth in this text are in accord with current recommendations and practice at the time of publication. However, in view of ongoing research, changes in government regulations, and the constant flow of information relating to drug therapy and drug reactions, the reader is urged to check the package insert for each drug for any change in indications and dosage and for added warnings and precautions. This is particularly important when the recommended agent is a new and/or infrequently employed drug.

P.O. Box, CH–4009 Basel (Switzerland)
Printed in Switzerland on acid-free paper by
Reinhardt Druck, Basel
ISBN 3–8055–6987–4

KARGER

Fax+ 41 61 306 12 34
E-Mail karger@karger.ch
www.karger.com

Vol. 48, Suppl. 1, 1999

Gynecologic and
Obstetric Investigation

Contents

1 **Introduction**
Hoshiai, H. (Osaka-Sayama)

Session 1
Reviews

2 **Paracrine Regulation of Matrix Metalloproteinase Expression in the Normal Human Endometrium**
Osteen, K.G.; Keller, N.R.; Feltus, F.A.; Melner, M.H. (Nashville, Tenn.)

14 **Angiogenesis in Endometriosis and Angiogenic Factors**
Fujimoto, J.; Sakaguchi, H.; Hirose, R.; Wen, H.; Tamaya, T. (Gifu)

21 **Expression of Aromatase Cytochrome P450 in Eutopic Endometrium and Its Application as a Diagnostic Test for Endometriosis**
Kitawaki, J.; Kusuki, I.; Koshiba, H.; Tsukamoto, K.; Honjo, H. (Kyoto)

Session 2
Reviews

29 **Endometriosis and Free Radicals**
Ota, H. (Akita); Igarashi, S. (Hachirogata); Hatazawa, J.; Tanaka, T. (Akita)

36 **Immunohistochemical Study of Angiogenic Factors in Endometrium and Endometriosis**
Fujishita, A.; Hasuo, A.; Khan, K.N.; Masuzaki, H.; Nakashima, H.; Ishimaru, T. (Nagasaki)

Session 3
Review

45 **The Pontential Role of Environmental Toxins in the Pathophysiology of Endometriosis**
Bruner-Tran, K.L.; Rier, S.E.; Eisenberg, E.; Osteen, K.G. (Nashville, Tenn.)

57 **Summary**
Taketani, Y. (Tokyo)

KARGER

Fax+ 41 61 306 12 34
E-Mail karger@karger.ch
www.karger.com

Access to full text and tables of contents, including tentative ones for forthcoming issues:
www.karger.com/journals/goi/goi_bk.htm

Gynecologic and
Obstetric Investigation

Gynecol Obstet Invest 1999;48(suppl 1):1

Introduction

Hiroshi Hoshiai

This supplement presents the latest research in endometriosis which was featured at the 2nd Japan Conference on Endometriosis, held in Kanagawa, Japan at the Hakone Prince Hotel on May 8, 1999, and sponsored by Takeda Pharmaceutical Industries, Ltd. Attendees at this conference were fortunate to hear informative presentations and stimulating discussions from many bright investigators at the forefront of this field.

Professor Kevin Osteen of Vanderbilt University School of Medicine, USA, discussed the regulation of endometrial matrix metalloproteinase as a possible therapeutic strategy for the treatment of endometriosis. He used in vitro models to investigate the relationships between progesterone-associated cytokines and matrix metalloproteinase expression. In addition, in a second talk, Dr. Osteen presented evidence of concern to us all about a possible association between endometriosis and environmental exposure to dioxin. His group investigated whether dioxin can disrupt steroid regulation of endometrial matrix metalloproteinases.

Angiogenesis in the normal uterine endometrium and in tumors occurs at a rapid rate. Assistant Professor Jiro Fujimoto of Gifu University, Japan, reported on angiogenesis in endometriosis, and discussed how the various angiogenic factors may contribute to vascular pathogenesis of endometriosis under the influence of ovarian steroids.

Assistant Professor Jo Kitawaki of Kyoto Prefectural University of Medicine, Japan, evaluated the clinical usefulness of a diagnostic test for endometriosis based on the detection of an aromatase cytochrome, P450. This new technique may allow clinicians to screen for the presence of endometriosis at outpatient infertility clinics.

It has long been known that free radicals are critical to the proper functioning of the reproductive process. Assistant Professor Hirotaka Ota of Akita University School of Medicine, Japan, shared his findings concerning the role of free radicals in the process of implantation, and the effect of superoxide dismutase expression in endometriosis.

Recently, new criteria for pigmented lesions in pelvic endometriosis have been established, and lesions are now macroscopically classified as red, black, or white. Dr. Akira Fujishita of Nagasaki University reported the latest findings on immunohistochemical studies to investigate differences in these pigmented lesions according to proliferation or expression of angiogenic factors.

All the investigators featured in this supplement are on the cutting edge of research on endometriosis. New understanding of morphologic and biochemical molecular function, such as that distinguishing eutopic and ectopic endometrial tissue, are critical to implementing new therapies for the successful treatment of endometriosis. We are fortunate to share in the discoveries presented in this supplement to advance the knowledge base on this serious disease.

KARGER

Fax + 41 61 306 12 34
E-Mail karger@karger.ch
www.karger.com

Accessible online at:
www.karger.com/journals/goi

Session 1
Review

Gynecologic and
Obstetric Investigation

Gynecol Obstet Invest 1999;48(suppl 1):2–13

Paracrine Regulation of Matrix Metalloproteinase Expression in the Normal Human Endometrium

Kevin G. Osteen Nancy R. Keller F. Alex Feltus Michael H. Melner

Women's Reproductive Health Research Center, Vanderbilt University School of Medicine, Nashville, Tenn., USA

Key Words
Progesterone · Cytokines · Endometrium · MMPs

Abstract
Endometrial expression of matrix metalloproteinase (MMP)-3, MMP-7 and MMP-11 occurs during menstrual breakdown and subsequent estrogen-mediated growth, but not during the secretory phase. These enzymes are suppressed by progesterone treatment. Paracrine factors, including transforming growth factor-β (TGF-β) and retinoic acid, are also critical for MMP regulation in the endometrium. In contrast, inflammatory cytokines such as interleukin-1α may block or interfere with steroid-mediated MMP regulation at ectopic sites of growth. Using in vitro models, our laboratory has investigated the complex interactions between progesterone and locally produced cytokines that may affect MMP expression during the development of endometriosis. Our results indicate that targeting the regulation of MMPs may represent an appropriate therapeutic strategy for the treatment of endometriosis.

The Pathophysiology of Endometriosis

The disease endometriosis is defined as the growth of functional endometrial glands and stroma at an extra-uterine site, most frequently within the peritoneal cavity. First identified as a disease over a hundred years ago [1], the cause of endometriosis continues to be debated and several hypotheses have been advanced. Nevertheless, most medical practitioners agree that incidences of the disease can be linked to the ectopic implantation of extra-uterine endometrial tissue entering the peritoneal cavity at the time of menstruation [2, 3]. Accurate diagnosis of endometriosis is complex, requiring invasive surgical techniques or expensive instrumentation that is not universally available. Therefore, the exact incidence of the disease is not known, but endometriosis is reported to affect 10–15% of American women of reproductive age and as many as 35% of infertile women [4]. In some countries, such as Belgium, the incidence of endometriosis has been reported to occur at a much higher frequency [5]. Endometriosis is a persistent disease process and complications such as bowel obstruction, inflammation, or excessive bleeding can become life threatening. Although malignant transformation of ectopic endometrial lesions occurs only rarely, the cellular mechanisms required for establishment of ectopic endometrial growth represent invasive events similar to those seen in neoplastic metastasis, and involve extensive degradation of the extracellular matrix (ECM) [6].

KARGER
Fax + 41 61 306 12 34
E-Mail karger@karger.ch
www.karger.com

0378–7346/99/0485–0002$17.50/0

Accessible online at:
www.karger.com/journals/goi

Kevin G. Osteen, PhD, Director
Women's Reproductive Health Research Center, B 1100 Medical Center North
Department of Obstetrics and Gynecology, Vanderbilt University School of Medicine
Nashville, TN 37232 (USA)
Tel. +1 615 322 4196, Fax +1 615 343 7913, E-Mail kevin.osteen@mcmail.vanderbilt.edu

Endometrial Remodeling during the Menstrual Cycle

The human endometrium is a unique adult tissue that undergoes a development-like process leading to regrowth of a functional organ system during each menstrual cycle. Matrix metalloproteinases (MMPs) are expressed in the endometrium in association with tissue growth and remodeling, as well as during tissue breakdown and sloughing at the end of the menstrual cycle [7, 8]. The MMPs are a large family of related enzymes that are the normal mediators of ECM degradation. Numerous MMPs have been identified in reproductive tissues during cellular processes that require the physical breakdown or restructuring of matrix proteins [9]. In the endometrium, several MMPs of the stromelysin family, including MMP-3, MMP-7, and MMP-11, are broadly expressed during menstruation and exhibit focal patterns of expression during repair and estrogen-mediated regrowth [7, 8]. Since MMPs are potentially destructive enzymes, their expression and activation is highly regulated within normal tissues. MMPs can be regulated transcriptionally as well as by posttranslation mechanisms via local tissue factors, including tissue inhibitors of MMPs (TIMPs), plasminogen activator inhibitor (PAI-1), and general protease inhibitors such as α-2 macroglobulin and α-1 antitrypsin [9–12]. Although local factors can regulate MMP activity in multiple tissues, the steroid sensitivity of the endometrium may provide a mechanism for a more dominate role for steroid influence on MMP expression in this tissue. Specifically, ovarian synthesis of estradiol induces growth and associated MMP expression, whereas ovarian progesterone serves as a potent inhibitor of MMP expression [7, 8]. In vitro observations have confirmed that progesterone exposure of isolated endometrial cells or organ cultures inhibits expression of MMP-3, MMP-7, and MMP-11 [13, 14].

Reflecting the steroid sensitivity of the endometrium, the development of endometriosis is a steroid-dependent process [15] and the disease appears more frequently in women after increased exposure to estrogens [16]. In contrast, a woman's exposure to progestins may limit development and progression of the disease [17, 18]. Although the precise cellular mechanisms that link steroid exposure to the disease process of endometriosis are not yet established, steroidal regulation of MMP expression may represent an important biological connection. A number of studies have shown expression of MMPs in ectopic endometrium [19, 20] and a decrease in their expression after successful medical therapy [19, 21]. We have more directly implicated MMPs in the establishment of endometriosis by showing that progesterone-mediated MMP suppression or therapeutic injections with TIMP-1, a natural MMP inhibitor, can prevent the establishment of ectopic human endometrial lesions in the peritoneum of nude mice [22]. It is tempting to speculate that suppression of MMPs may be linked to the therapeutic benefits often associated with pregnancy or progestin therapy in reducing or preventing endometriosis. However, many women fail to respond adequately to progestin therapy, and a diminished response to progesterone has been documented at ectopic sites of endometrial growth [23, 24]. Not surprisingly, there appears to be an altered expression of both MMP-3 and MMP-7 in endometriotic lesions compared to normal endometrial tissues [19, 20]. Even if regulation of MMP expression represents a critical process in the overall pathophysiology of endometriosis, it is unlikely that the action of progestins alone can effectively treat the diverse presentations and aggressive nature of this complex disease.

Cytokine Regulation of MMP Expression

A broad array of locally produced cytokines act within endometrial tissues to enhance or modify the impact of steroid exposure. The local expression of cytokines precisely regulate endometrial function, including cell-specific MMP expression during the cycle [14, 25, 26]. For example, both in vivo and in vitro findings have shown that progesterone-mediated suppression of endometrial MMPs involves both direct cellular effects and indirect effects mediated by secondary paracrine pathways [7, 8, 13, 27, 28]. Therefore, superimposed over the effects of steroid action are additional cytokine-associated signaling pathways that affect endometrial MMP expression. Unfortunately, because of such multiple interactive regulatory pathways, inadequate ovarian progesterone production, diminished tissue responsiveness to progesterone, or altered patterns of local cytokine expression can each disrupt normal MMP regulation. Because MMPs appear to contribute to the development of endometriosis, our research has focused on cytokines that act in concert with progesterone to limit MMP expression. To date, these studies have shown that progesterone and its associated cytokines suppress MMP gene and protein expression and also decrease the stimulatory actions that inflammatory cytokines can have on MMP expression.

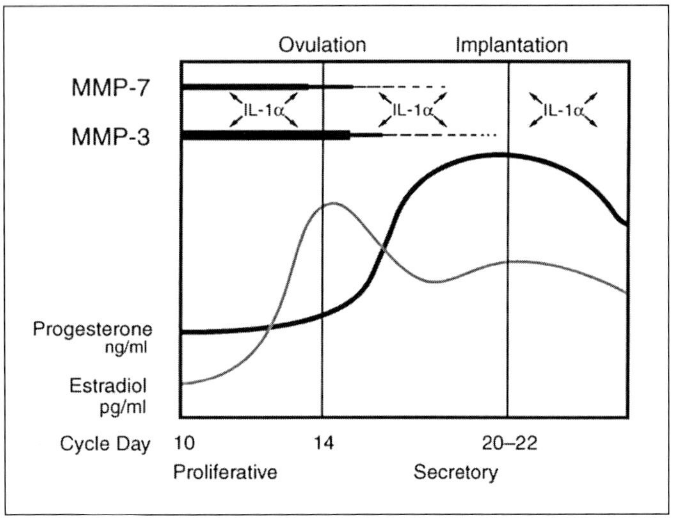

Fig. 1. Schematic representation of the effects of exogenous IL-1α on MMP-3 and MMP-7 expression by human endometrial organ cultures obtained throughout the menstrual cycle. After ovulation and the rise in serum progesterone, the ability of IL-1α to stimulate MMP expression is greatly diminished.

Interleukin-1

The interleukin-1 (IL-1) family consists of IL-1α, IL-1β, IL-1 receptor antagonist (IL-1ra), and the receptors IL-1 RtI and IL-1 RtII [29, 30]. Several laboratories have shown that exposure to either IL-1α or IL-1β stimulates production of MMP-1, MMP-3, and MMP-9 by endometrial stromal cells [31–34] and MMP-7 in epithelial cells [34, 35]. Curiously, although in vitro IL-1α cotreatment can prevent progesterone suppression of MMPs in proliferative phase endometrial organ cultures [34, 36], immunolocalization studies have shown that IL-1α is expressed in vivo during the secretory phase [26, 37, 38] when MMPs are suppressed. Singer et al. [32] showed that progesterone can inhibit both the release of IL-1α by endometrial epithelial cells and stromal cell expression of MMP-1 in response to this cytokine. We have found that while endometrial tissues removed during the proliferative phase produce MMP-3 and MMP-7 in response to IL-1α, tissues removed during the secretory phase are resistant to MMP stimulation by this cytokine (summarized in figure 1) [36]. In addition, IL-1α is unable to stimulate MMP-3 expression in endometrial stromal cells exposed to progesterone for as little as two days [Osteen, unpublished data, 1999] supporting a role for this steroid in the diminished endometrial sensitivity to IL-1α. Just prior to menstruation, endometrial MMPs are re-expressed as progesterone levels fall [7, 8, 39] and epithelial cells release IL-1α [32]. Taken together, these studies would indicate that progesterone, in promoting cellular differentiation, also promotes a poorly understood resistance within endometrial cells to the affects of local inflammatory cytokines.

Given the complexity of steroid and cytokine-mediated MMP regulation, it is not surprising that many women with endometriosis fail to respond significantly to steroid therapy alone [20]. In addition, IL-1 has been reported to be elevated in the peritoneal fluid of women with endometriosis [40] and may stimulate the secretion of MMPs during ectopic implantation of endometrial tissue. Our studies have found that IL-1α can stimulate MMP expression in endometrial tissue fragments, opposing the normally suppressive action of progesterone, in an experimental endometriosis model [34, 41]. However, as noted above, progestin therapy can be successful in the treatment of endometriosis and, under experimental conditions, progesterone can limit the effects of locally produced inflammatory cytokines. Identifying which local tissue factors can act in concert with progesterone to suppress MMPs in opposition to inflammatory cytokines should provide additional therapeutic opportunities in the treatment of endometriosis.

Transforming Growth Factor-β

In numerous biological systems, members of the transforming growth factor-β (TGF-β) family are known to enhance deposition of ECM components [42] and inhibit expression of many MMPs [43–45]. Although TGF-β is present in the human endometrium throughout the menstrual cycle, an increased expression of this cytokine occurs during the secretory phase and in early pregnancy [46]. In vitro cocultures of purified populations of human endometrial epithelial and stromal cells showed TGF-β as a key progesterone-dependent factor, necessary for suppression of MMP-7 [28]. However, the mechanism by which progesterone promotes the increased expression of a specific TGF-β isotype is not known. Interestingly, the human TGF-β_2 promoter does not contain a traditional progesterone response element (PRE), a requirement for certain well-characterized interactions between steroid receptors and DNA. Recently, investigators have suggested that the progesterone receptor may be able to activate transcription of specific genes by interacting with other transcriptional factors [47, 48]. In a preliminary investigation, our group examined whether the progesterone

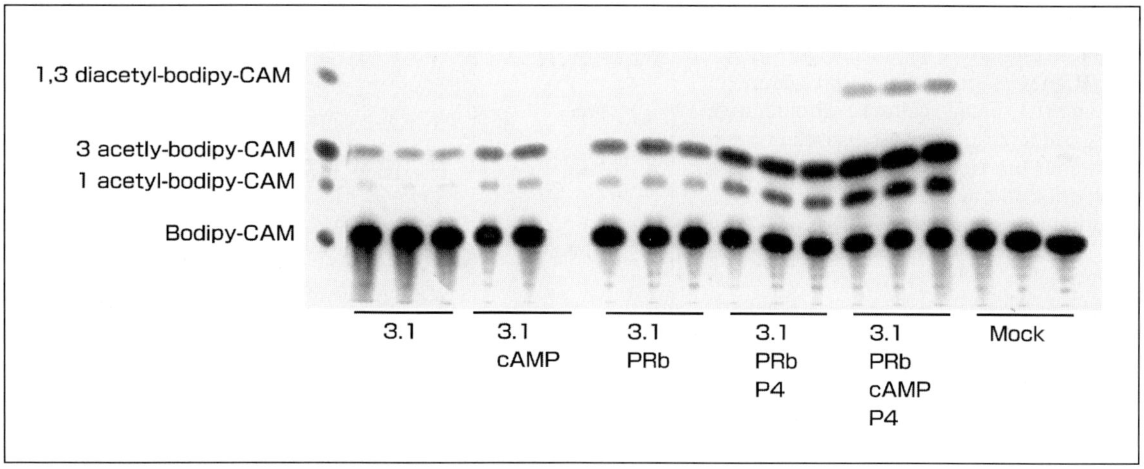

Fig. 2. Thin layer chromatography plate of CAT enzyme activity. Hela cell extracts were obtained from cells transfected with full-length (–3130 to +73) TGF-β_2 promoter-CAT construct (3.1) in the presence of the B form of the progesterone receptor (PRb). The cells were transfected, glycerol-shocked, washed and treated for 24 h in the presence of or absence of 100 n*M* progesterone (P4), 0.5 m*M* cAMP (cAMP) or both progesterone and cAMP. The cells were collected and cell extracts prepared and assayed with fluorescent BODIPY-CAM substrate as described elsewhere [54]. The image is a scanned negative of a plate illuminated with 366 nm direct ultraviolet light. Each lane represents extracts from a separate transfected culture dish of cells performed in triplicate. Cells from lane 7 were lost due to technical error.

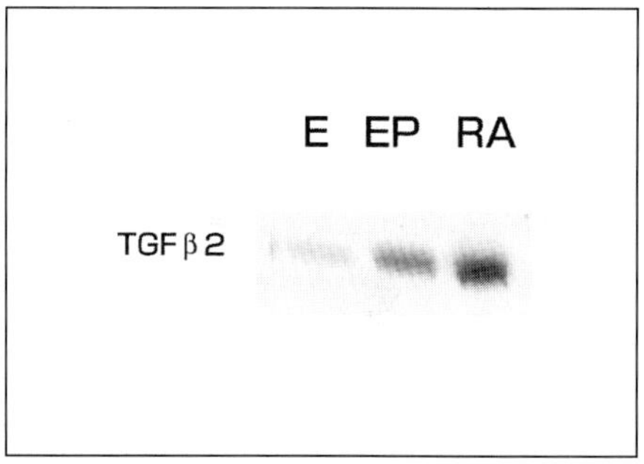

Fig. 3. Western blot analysis of TGF-β_2 expression and regulation in endometrial organ cultures. Proliferative phase endometrium was established in organ culture as previously described [28] and treated for 48 h with estradiol (E, 10^{-8} *M*), estradiol plus progesterone (EP, E = 10^{-9} *M*; P = 5×10^{-7} *M*) or estradiol plus retinoic acid (RA; E = 10^{-9} *M*; RA = 10^{-6} *M*). Western blot analysis was performed as previously described [41].

receptor may affect the TGF-β_2 promoter via a novel, nontraditional mechanism. For this study, Hela cells were co-transfected with the progesterone receptor and a 3.1-kb piece of the TGF-β_2 promoter fused to a chloramphenicol acetyl transferase (CAT) reporter construct. As shown in figure 2, although CAT activity was only minimally stimulated by progesterone or cAMP alone, enzyme activity could be greatly stimulated by co-addition of these agents. Although other investigators have shown that TGF-β_2 gene expression is increased by cAMP or forskolin [49], our results may be the first to indicate a possible interaction of the progesterone receptor with the TGF-β_2 promoter.

Whether or not further evidence indicates a direct effect of progesterone on TGF-β_2 expression, it is unlikely that progesterone action alone accounts for the increased TGF-β_2 expression that is observed during endometrial maturation. In numerous biological systems, retinoic acid, a vitamin A derivative, serves an important role in the regulation of TGF-β expression and action. For example, retinoic acid has been shown to specifically enhance TGF-β_2 expression [50, 51], stimulate posttranscriptional activation and enhance cellular responses to TGF-β_2 [51]. It is likely that retinoic acid will be found to play an equally important role in regulating TGF-β expression or action in the endometrium. We have recently found that

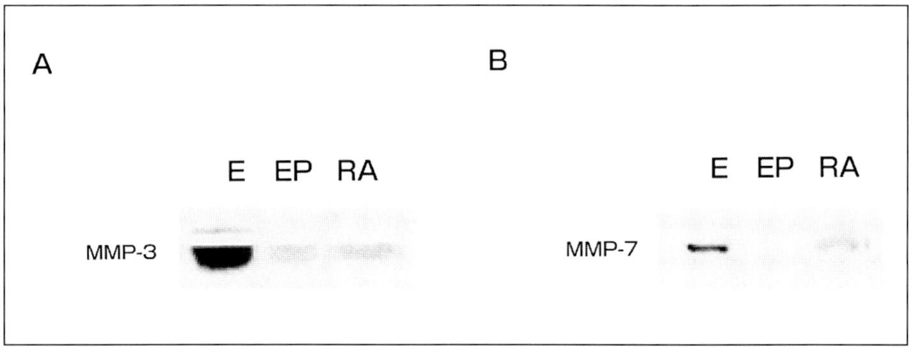

Fig. 4. Western blot analysis of (**A**) MMP-3 and (**B**) MMP-7 expression and regulation in endometrial organ cultures. Proliferative phase endometrium was established in organ culture as previously described [28] and treated for 48 h with estradiol (E, 10^{-8} *M*), estradiol plus progesterone (EP, E = 10^{-9} *M*; P = 5 × 10^{-7} *M*) or estradiol plus retinoic acid (RA; E = 10^{-9} *M*; RA = 10^{-6} *M*). Western blot analysis was performed as previously described [41].

local retinoic acid synthesis is increased more than 100-fold by progesterone in an in vitro decidualization model [Osteen, unpublished data, 1999]. As shown in figure 3, retinoic acid treatments alone, in a fashion similar to progesterone treatment, can stimulate the secretion of TGF-β_2 in short-term endometrial organ cultures. Taken together, these studies indicate that the control of TGF-β expression and action by both progesterone and retinoic acid may represent an interactive pathway of critical importance in the endometrium.

Retinoic Acid

As noted earlier, our laboratory has shown that stromal cells, derived from human endometrium synthesize retinoic acid, an active metabolite of vitamin A, during the process of in vitro decidualization [Osteen, unpublished data, 1999]. It has been known for some time that female rats deficient in vitamin A display abnormal uterine morphology and subsequent pregnancy failure [52]. The capacity of endometrial stromal cells to synthesize retinoic acid appears to be dependent on exposure to progesterone, because endometrial stromal cells obtained from proliferative phase endometrium cannot produce significant amounts of retinoic acid in culture. Retinoic acid binds to specific members of the nuclear receptor family (RAR and RXR) and can regulate transcription in a manner similar to that of steroid hormones [53]. In addition to affecting TGF-β expression and action, retinoic acid may play an important role in the regulation of MMPs, a position supported by both in vitro and in vivo observations. As shown in figure 4, retinoic acid treatments can suppress the secretion of either MMP-3 or MMP-7 in organ cultures of human endometrium. Retinoic acid, like progesterone, can reduce the ability of IL-1 to stimulate MMP-3 secretion by isolated endometrial stromal cells [Osteen, unpublished data, 1999], further linking the action of retinoic acid to that of progesterone in the endometrium. While it may be tempting to speculate that retinoic acid could replace progesterone action in the suppression of MMPs in the secretory endometrium, it must be emphasized that progesterone is essential for decidualization and therefore for sustained retinoic acid production as well as TGF-β_2 expression. In addition, the secondary paracrine effects of TGF-β are critical for either progesterone or retinoic acid regulation of MMP expression. The addition of blocking antibodies to TGF-β will prevent MMP suppression in human endometrial organ cultures after either progesterone or retinoic acid treatments (fig. 5).

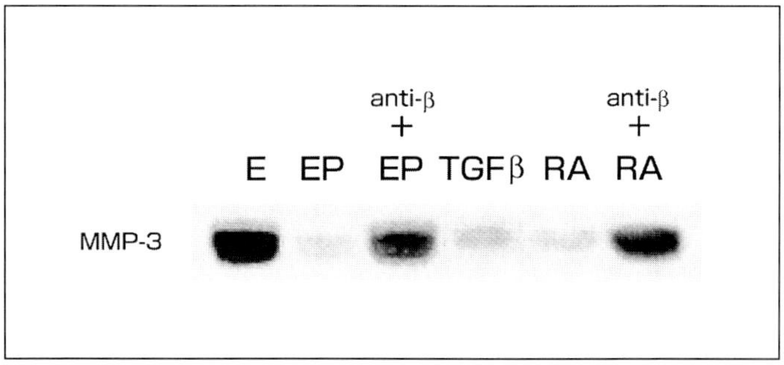

Fig. 5. Western blot analysis of MMP-3 expression and regulation in endometrial organ cultures. Proliferative phase endometrium was established in organ culture as previously described [28] and treated for 48 h with estradiol (E, 10^{-8} *M*), estradiol plus progesterone (EP, E = 10^{-8} *M*; P = 5 × 10^{-7} *M*), estradiol plus TGF-β (TGF-β; E = 10^{-9} *M*; TGF-β = 2 ng/ml), estradiol plus retinoic acid (RA; E = 10^{-8} *M*; RA = 10^{-6} *M*). Some cultures also received 10 ng/ml of a pan-specific antibody to TGF-β (anti-β). Western analysis was performed as previously described [41].

Summary

Laboratory and clinical experience would suggest that steroids contribute to the development and progression of endometriosis [4, 15, 16], although progesterone therapy and the endocrine environment of pregnancy may act to prevent or suppress the disease process [17, 18]. Unfortunately, many women fail to respond to progestins, possibly resulting from an insensitivity to this steroid [23, 24] or to the presence of local tissue inflammation, which may alter steroid sensitivity. In the normal endometrium, both direct and indirect steroid-mediated pathways have been shown to be necessary for MMP suppression [13, 27, 28], and local cytokines can impact the action of progesterone on the establishment of human endometrial lesions in an experimental model of endometriosis [34, 41]. Our laboratory has focused considerable attention toward better understanding the role of progesterone in the regulation of MMPs in the normal endometrium as well as in the disease endometriosis. By studying the process of decidualization in vitro [Osteen, unpublished data, 1999] we find that the local synthesis of retinoic acid may be an important component of progesterone action during the establishment of pregnancy [14]. In the current study, we have determined that retinoic acid can act in a fashion similar to progesterone in the regulation of endometrial MMPs. Moreover, retinoic acid action, like progesterone action, can be blocked by an antibody to TGF-β. Retinoic acid can decrease endometrial MMP-3 expression, inhibit stimulation of MMP-3, and reduce the establishment and progression of experimental endometriosis. Although little is known about the role of retinoic acid in the disease endometriosis at this time, this vitamin A derivative is known to be an essential component of normal endometrial function, including the maintenance of normal pregnancy. Developing a better understanding of MMP regulation by both steroids and cytokines should provide insight into developing more successful treatment strategies for endometriosis.

References

1 Von Rokitansky C: Über Uterusdrüsen-Neubildung im Uterus und Ovarialsarcomen. Zkk Gesellsch d Aerzte zu Wien 1860;37:577.

2 Sampson JA: Peritoneal endometriosis due to menstrual dissemination of endometrial tissue into the peritoneal cavity. Am J Obstet Gynecol 1927;14:422–469.

3 Ishimaru T, Masuzaki H: Peritoneal endometriosis: Endometrial tissue implantation as a primary etiologic mechanism. Am J Obstet Gynecol 1991;165:210–214.

4 Cramer DW, Wilson E, Stillman RJ, Berger MJ, Belisle S, Schiff I, Albrecht B, Gibson M, Stadel BV, Schoenbaum SC: The relationship of endometriosis to menstrual characteristics, smoking and exercise. J Am Med Assoc 1986; 255:1904–1908.

5 Koninckx PR, Braet P, Kennedy SH, Barlow DH: Dioxin pollution and endometriosis in Belgium. Hum Reprod 1994;9:1001–1002.

6 Spuijbroek MDEH, Dunselman GAJ, Menheere PPCA, Evers JLH: Early endometriosis invades the extracellular matrix. Fertil Steril 1992;58:929–933.

7 Rodgers WH, Osteen KG, Matrisian LM, Navre M, Gorstein F: Expression and localization of matrilysin, a matrix metalloproteinase, in the human endometrium. Am J Obstet Gynecol 1993;168:253–260.

8 Rodgers WH, Matrisian LM, Giudice LC, Dsupin B, Cannon P, Svitek C, Gorstein F, Osteen KG: Patterns of matrix metalloproteinase expression in cycling endometrium imply differential functions and regulation by steroids. J Clin Invest 1994;94:946–953.

9 Hulboy DL, Rudolph LA, Matrisian LM: Matrix metalloproteinases as mediators of reproductive function. Mol Hum Reprod 1997;3: 27–45.

10 Lockwood CJ, Schatz F: A biological model for the regulation of peri-implantational hemostasis and menstruation. J Soc Gynecol Invest 1996;3:159–165.

11 Oleksyszyn J, Augustine AJ: Plasminogen modulation of IL-1-stimulated degradation in bovine and human articular cartilage explants. The role of the endogenous inhibitors: PAI-1, α_1-PI, α_2-macroglobulin and TIMP. Inflamm Res 1996;45:464–472.

12 Woessner JFJ: Regulation of matrilysin in the rat uterus. Biochem Cell Biol 1996;74:777–784.

13 Schatz F, Papp C, Toth-pal E, Lockwood C: Ovarian-steroid modulation of stromelysin-1 expression in human endometrial stroma and decidual cells. J Clin Endocrin Metab 1994;78: 1467–1472.

14 Osteen KG: The endocrinology of uterine decidualization; in Bazer FW (ed): The Endocrinology of Pregnancy. Totowa, Humana Press, 1998, pp 543–552.

15 DiZerga GS, Barber DL, Hodgen GD: Endometriosis: Role of steroids in initiation, development, maintenance and suppression. Fertil Steril 1980;33:649–653.

16 Halme J, Stovall D. Endometriosis and its medical management; in Wallach EE, Zacur HA (eds): Reproductive Medicine and Surgery. St. Louis, Mosby, 1995, pp 695–710.

17 Ramzy, I: Pathology; in Schenken RS (ed): Endometriosis: Contemporary Concepts and Clinical Management. Philadelphia, Lippincott, 1989, p 60.

18 Olive DL: Medical treatment: Alternatives to danazol; in Schenken RS (ed): Endometriosis: Contempory Concepts and Clinical Management. Philadelphia, Lippincott, 1989, pp 192–202.

19 Saito T, Mizumoto H, Kuroki K, Fujii M, Mori S, Kudo R: Expression of MMP-3 and TIMP-1 in endometriosis and the influence of Danazol. Acta Obstet Gynecol Japan 1995;47:495–496.

20 Osteen KG, Bruner KL, Sharpe-Timms, KL: Steroid and growth factor regulation of matrix metalloproteinase expression and endometriosis. Sem Rep Endocr 1996;14:247–255.

21 Sharpe-Timms KL, Keisler LW, McIntush EW, Keisler DH: Tissue inhibitor of metalloproteinase-1 concentrations are attenuated in peritoneal fluid and sera of women with endometriosis and restored in sera by gonadotropin releasing hormone-agonist therapy. Fertil Steril 1998;69:1128–1134.

22 Bruner KL, Matrisian, LM, Rodgers WH, Gorstein F, Osteen KG: Suppression of matrix metalloproteinases inhibits establishment of ectopic lesions by human tissue in nude mice. J Clin Invest 1997;99:2851–2857.

23 Lessey BA, Metzger DA, Haney AF, McCarty KS Jr: Immunohistochemical analysis of estrogen and progesterone receptors in endometriosis: Comparison with normal endometrium during the menstrual cycle and the effect of medical therapy. Fertil Steril 1989;49:229–235.

24 Bergqvist A, Ferno M: Oestrogen and progesterone receptors in endometriotic tissue and endometrium: comparison of different cycle phases and ages. Hum Reprod 1993;8:2211–2217.
25 Giudice LC: Growth factors and growth modulators in human uterine endometrium: Their potential relevance to reproductive medicine. Fertil Steril 1994;61:1–7.
26 Tabibzadeh S, Sun XZ. Cytokine expression in human endometrium throughout the menstrual cycle. Hum Reprod 1992;7:1214–1221.
27 Osteen KG, Rodgers WH, Gaire M, Hargrove J, Gorstein F, Matrisian LM: Stromal-epithelial interaction mediates steroidal regulation of matrix metalloproteinase expression in the human endometrium. Proc Natl Acad Sci USA 1994;91:10129–10133.
28 Bruner KL, Rodgers WH, Korc M, Gold L, Hargrove J, Matrisian LM Osteen KG: Transforming growth factor-β mediates the progesterone suppression of an epithelial metalloproteinase by adjacent stroma. Proc Natl Acad Sci USA 1995;92:7362–7366.
29 Dinarello CA, Wolff SM: The role of Interleukin-1 in disease. New Engl J Med 1993;328: 106–113.
30 Arend WP: Interleukin-1 receptor antagonist. Adv Immunol 1993;54:167–227.
31 Rawdanowicz TJ, Hampton AL, Nagase H, Woolley DE, Salamonsen LA: Matrix metalloproteinase production by cultured human endometrial stromal cells: Identification of interstitial collagenase, gelatinase A, gelatinase B, and stromelysin-1 and their differential regulation by interleukin-1. Biochem J 1994;79:530–536.
32 Singer CF, Marbaix E, Kokorine I, Lemoine P, Donnez J, Eeckhout Y, Courtoy PJ: Paracrine stimulation of interstitial collagenase (MMP-1) in the human endometrium by interleukin 1 alpha and its dual block by ovarian steroids. Proc Natl Acad Sci USA 1997;94:10341–10345.
33 Huang HY, Wen Y, Irwin JC, Kruessel JS, Soong YK, Polan ML: Cytokine-mediated regulation of 92-kilodalton Type IV collagenase, tissue inhibitor of metalloproteinase-1 (TIMP-1), and TIMP-3 messenger ribonucleic acid expression in human endometrial stromal cells. J Clin Endocrinol Metab 1998;83:1721–1729.
34 Bruner KL, Keller NR, Osteen KG: Interleukin-1α opposes suppression of human endometrial matrix metalloproteinases by progesterone in a model of experimental endometriosis. Proceedings of the VI World Congress on Endometriosis, 1998.
35 Osteen KG, Sierra-Rivera E, Keller NR, Fox D: Interleukin-1α opposes progesterone-mediated suppression of MMP-7: A possible role of this cytokine in human implantation. Annal NY Acad Sci 1997;828:137–145.
36 Osteen KG, Sierra-Rivera, Keller NR, Boehm FH, Eisenberg E: Progesterone-mediated suppression of matrix metalloproteinases is opposed by interleukin-1α during human implantation. Biol Rep 1997;56:176. (abstr.).
37 Simon C, Piquette GN, Frances A, Polan ML: Localization of interleukin-1 Type I receptor and interleukin-1 in human endometrium throughout the menstrual cycle. J Clin Endocrinol Metab 1993;77:549–555.
38 Simon C, Mercader A, Frances A, Gimeno MJ, Polan ML, Remohi J, Pellicer A: Hormonal regulation of serum and endometrial IL-1 alpha, IL-1 beta and IL-1ra: IL-1 endometrial microenvironment of the human embryo at the apposition phase under physiological and supraphysiological steroid level conditions. J Reprod Immunol 1996;31:165–184.
39 Lockwood CJ, Krikun G, Hausknecht VA, Papp C, Schatz F: Matrix metalloproteinase and matrix metalloproteinase inhibitor expression in endometrial stromal cells during progestin-initiated decidualization and menstruation-related progestin withdrawal. Endocrinol 1998;139:4607–4613.
40 Fakih H, Bagget B, Holtz G, Tsang KY, Lee JC, Williamson HO: Interleukin-1: Possible role in the infertility associated with endometriosis. Fertil Steril 1987;47:213–217.
41 Bruner KL, Eisenberg E, Gorstein F, Osteen KG: Progesterone and Transforming growth factor-β coordinately regulate suppression of endometrial matrix metalloproteinases and prevention of experimental endometriosis. Steroids 1999;64:648–653.
42 Roberts AB, McCune BK, Sporn MB: TGF-β: Regulation of extracellular matrix. Kidney Int 1992;41:557–559.
43 Edwards DR, Murphy G, Reynolds JJ, Whitham SE, Docherty AJP, Angel P, Health JK: TGF-β modulates the expression of collagenase and metalloproteinase inhibitor. EMBO J 1987;6:189–194.
44 Kerr L, Miller D, Matrisian LM: TGF-β 1 inhibition of transin/stromelysin-1 gene expression is mediated through a fos binding sequence. Cell 1990;61:267–278.
45 Marti HP, Lee L, Kashgarian M, Lovett DH: TGF-β_1 stimulates glomerular mesangial cell synthesis of the 72 kDa Type IV collagenase. Am J Path 1994;144:82–94.
46 Kauma S, Matt D, Strom S, Eirman D, Turner T: Interleukin-1β, human leukocyte antigen HLA-DRα, and transforming growth factor-β expression in endometrium, placenta, and placental membranes. Am J Obstet Gynecol 1990; 163:1430–1437.
47 Handel EG, Melner MH, Feltus FA, Edwards DP: Progesterone negatively regulates the β-casein promoter through a composite progesterone receptor-Stat5 element (abstr.). Presented at the 81st Annual Meeting of the Endocrine Society, San Diego, 1999.
48 Feltus FA, Groner B, Edwards DP, Melner MH: Functional interaction of the progesterone receptor and Stat5 in the stimulation of the human type II 3β-hydroxysteroid dehydrogenase promoter: Evidence for an autocrine loop in progesterone biosynthesis (abstr.). Presented at the 81st Annual Meeting of the Endocrine Society, San Diego, 1999.
49 Roberts AB, Kim S-J, Noma T, Glick AB, Lafyatis R, Lechlieder R, Jakowlew SB, Geiser A, O'Reilly MA, Danielpour D, Sporn MB: Clincal applications of TGF-β. Ciba Foundation Symposium 1991;157:7–28.
50 Glick AB, Flanders KC, Danielpour D, Yuspa SH, Sporn MB: Retinoic acid induces transforming growth factor-β_2 in cultured keratinocytes and mouse epidermis. Cell Reg 1989;1: 87–89.
51 Nugent P, Linh MA, Greene RM: Differential expression and biological activity of retinoic acid-induced TGF isoforms in embryonic palate mesenchymal cells. J Cell Physiol 1998;177: 36–46.
52 Thompson JN, MacHowell J, Pitt GAJ: Vitamin A and reproduction in rats. Proc R Soc Lond Ser B 1964;159:510.
53 Giguere V: Retinoic acid receptors and cellular retinoid binding proteins: Complex interplay in retinoid signaling. Endo Rev 1994;15:61–97.
54 Young SL, Barbera L, Kaynard AH, Haugland RP, Kang HC, Brinkley M, Melner MH: A nonradioactive assay for chloramphenicol acetyltransferase activity using fluorescent substrates. Anal Biochem 1991;197:401–407.

Post-Presentation Discussion

Professor Terakawa: Thank you for a beautiful presentation. The paper is open for discussion please. By the way, you focus, Professor Osteen, on relation on expression of stromelysin, MMP-3 and epithelial specific MMP-7, matrilysin. There are 10 or 11 enzymes of MMPs. My question is, you focus on MMP-3 and MMP-7 and why not several other enzymes that are in the endometrial tissue? And, to my knowledge, and I was impressed when I read the article published by a Belgian group, I think, in 1992 in the Proceedings of the National Academy of Science, they reported progesterone prevented endometrial tissue breakdown by inhibiting MMP enzyme. In that lab, I remember the enzymes were gelatinase B, I think, and also collagenase, and yet, in your presentation, you didn't mention those enzymes. So, would you explain the focus on MMP-3 and MMP-7 rather than the other enzymes.

Professor Osteen: The answer to your question is simple in terms of our focus. Our focus is to understand how progesterone mediates MMP gene expression in the endometrium. We are maintaining a focus on a smaller number of MMPs that we have more specific knowledge about so that we can pursue the question of 'How are MMPs regulated by progesterone?' which is the important question that we feel we can lend an answer to, as opposed to a broader view of all the different MMPs that are involved in endometrium. There are a number of groups that are doing beautiful work, I think, on expanding the number of MMPs that have been implicated in the endometrium and endometriosis. I think that is important work. Our focus, has been actually to use the MMPs as markers of progesterone action, so that we can understand 'How does progesterone create a suppressive environment that inhibits multiple MMPs?'. So, we are focusing not so much on the broad range of MMP expression as on the specific question of how progesterone is acting through cell-cell communication to regulate these enzymes. So, we have a more narrow question that we are trying to answer.

Professor Terakawa: So, in other words, do other enzymes such as gelatinase and collagenase not alter and do not change during the menstrual cycle?

Professor Osteen: There are less changes in other MMPs across the menstrual cycles reported by some groups than the specific ones we are looking at. MMP-11, for example, may be more highly regulated in a wound response than MMP-3. We have some evidence that not all of the MMPs are as highly regulated as others are. I think that is a very interesting aspect of MMP action. And so, I think we will find that certain of the MMPs, such as the gelatinases may be not as highly regulated as other members of the family. It is not clear exactly why these differences occur, and why so many MMPs are expressed is not even understood. There is a lot of redundancy in the MMP system apparently. And so we are not sure what the answer will be in terms of how these different enzymes are used across the cycle and what the impact on inhibiting any specific enzyme might be on the disease process. So I think that is a very important question as we begin to look at which of the enzymes might be more involved in the disease process. The regulation of multiple MMPs is a different question than the one we are asking, but a very important question.

Professor Terakawa: One more question at this time, you mentioned the enzyme changes in the normal human endometrium. Have you seen such an alteration of changing of MMP-3 and MMP-7 in endometriosis?

Professor Osteen: What you find in the endometriotic lesion is there is a rather constant low-level expression of each of these enzymes, MMP-3, MMP-11, and MMP-7, that we have specifically looked at. So, the effect of progesterone on the normal endometrium is not being duplicated out in the peritoneal cavity, and this data has driven us to try to understand what out there is missing in terms of either cell-cell communication parameters or specific cytokines that might be misexpressed in the peritoneal cavity. Our hypothesis, of course, is that the cytokine environment in the peritoneal cavity, including interleukin and other factors that can regulate MMPs such as TNF-α. These factors are all highly up-regulated in the peritoneal cavity and they interfere with the normal ability that progesterone would have on suppressing MMPs. And again, we know from the literature that progesterone doesn't always act in the ectopic site the way it does in the normal endometrium. So there is something out there, there are many things in the peritoneal environment that are wrong, but we still have to develop a therapeutic approach. Our therapeutic approach is to go beyond progesterone to a cytokine that works in concert with progesterone to perhaps have a greater impact on tissue growth at the ectopic site. I don't think progesterone would be the answer to treating all endometriosis. I don't think any of us believe it is. We are going to have to come up with something that is more effective in reducing this disease impact.

Professor Kanzaki: Thank you very much for that very interesting presentation. I would like to ask one question as to your hypothesis as to the suppression of the MMP-7 by progesterone-induced TGF-β, maybe β_2, released from the stroma cell.

Professor Osteen: It is also, TGF-β_2 appears to be made during the differentiation of the stromal cell. This factor is also synthesized quite a bit by epithelial cells.

Dr. Kanzaki: In that case, TGF-β is usually produced by the cells in inactive form. Some substances which activate TGF-β are necessary for the localization of the active TGF-β. How do you explain the mechanism of the activations of TGF-βs? What kind of substance is necessary for the activation of TGF-β in the endometrium?

Professor Osteen: Well, again, retinoic acid may play a role in regulation of TGF-β activation. But, there are other factors. TGF-β can also bind to the matrix and be activated to a certain degree by the plasminogen activator systems at the invasive site. So, TGF-β in some way may be involved in the matrix activity that shuts down MMPs at the end of wound responses, whereas factors such as progesterone or retinoic acid might be acting more directly on the gene. Part of what I haven't talked about today is some of our work at looking at how progesterone might create such an inhibitory environment at the level of the MMP genes. But, I think that is where the key issue is going to come in. TGF-βs are such a complex family of growth factors. Many labs have become very frustrated trying to understand TGF-β action on any cell in any particular environment. So, we have sort of moved away from thinking that TGF-β-directed strategy would work because of the complex number of activators that are required for the TGF-β system to work. So, I am not sure what out in the peritoneum might be active or inactive that would cause TGF-β to be activated or not at the ectopic sites, but there is quite a bit of TGF-β expression in ectopic sites of endometrial growth, but it does not appear to be working very well to suppress the MMPs. So, there must be some factor other than TGF-β that is missing out in the peritoneal cavity.

Dr. Harada: Thank you for your nice presentation. I have two questions. One is about implantation. It has been published that interleukin 1 receptor is very important in mouse implantation. That issue is still controversial, but what do you think about that the importance of interleukin 1 in implantation? It also affects MMPs during the implantation period.

Professor Osteen: Yes. I have a curious hypothesis about that. The mouse work has remained controversial. To me, the way that I look at the interleukin system in the endometrium is that all of the pathways that turn MMPs on at a very high level exist and are actually enhanced in the secretory endometrium including the interleukin system. Invading macrophages are capable of making more of these cytokines. And, what you have is an environment which is in delicate balance so that when progesterone is driving a normal endometrial response, these cytokines are held in abeyance and don't act to stimulate MMPs. But, it also gives you a system in which, if progesterone levels drop, or if something about the pregnancy begins to fail, you have almost a self-destruct system that allows the endometrium to very rapidly disintegrate and expel products of conception. I think that could have a lot of benefit in failed pregnancy. Now, whether or not that same system is utilized during human implantation and whether the local environment of IL-1 is so high that you can stimulate MMPs, I am not sure. We have looked at ectopic pregnancy and found very high levels of MMP-7 at sites of ectopic pregnancy, but not in the normal endometrium at the same time. It is very difficult to get early sites of implantation in the human, so I can't answer the question about which MMPs might actually be turned on very focally. Our impression is that the IL-1 system is not very much able to stimulate endometrial MMPs as long as the endometrium is under the influence of progesterone. But, if you block progesterone with an abortifacient such as RU-486, you would have a rapid breakdown of that system. You would turn on the MMPs and you would abort. So, I think it could be important, but we have no evidence that the IL-1 system is regulating endometrial MMPs at the site of implantation.

Dr. Harada: Thank you. My next question is about endometriosis. We have been using stroma cells from ovarian endometrial tissues for experiment and we use the method you published in Fertility and Sterility in 1989. Many scientists suggest that there is a difference between eutopic, normal endometrium, and ectopic sites. For example, IL-6 expression, such a difference exists. And I would like to know as you illustrated, progesterone inhibits MMP-3 and MMP-11 expression in cultured stroma cells. What about ectopic sites? Stroma cells from ectopic sites?

Professor Osteen: It is a very good question. It is actually one of the specific aims of a new grant we have, to ask that very same question. Is the defect out at the ectopic site because the stromal cells at the ectopic site don't respond as much to progesterone? All I can say at this point is that we have mixed results on that question. It is not clear. Some stromal cells do respond to progesterone and some don't appear to. I don't know if our current results have to do with the population of stroma that we are getting. Again, if you have an endometrioma where you can get a much more pure population, it is much easier to work with. We have tried to look at some other sites in the peritoneum; but peritoneal lesions are more difficult to look at. But, it appears to be that progesterone has a variable effect on cells from ectopic sites. At this point, your lab may have more information on this subject than we do because again, we have focused on normal endometrial cells before we started moving out into regions where the normal control mechanisms would not be in place. So, if you have data on that area, I would like to hear it.

Dr. Harada: No, I don't have much data, but there is some difference between the peritoneal endometriosis and ovarian endometriosis.

Professor Osteen: Yes, the ovarian endometriosis appears to have more response to steroids than most of the peritoneal lesions. And, ovarian endometriosis is the most 'endometrium-like' of endometriosis. It is the best tissue to look at to make comparisons. It is much more difficult to study peritoneal lesions, which have so many different types of presentation. And, again, part of our new grant is to do these types of comparisons and we are doing them with Professor Robert Taylor, and you probably know his work as well at the University of California, San Francisco.

Dr. Furuya: Thank you very much. I am interested in the relationship between the implantation of the endometriotic tissue and normal peritoneal tissues concerning subperitoneal morphological or the biological change in patients with endometriosis. Have you ever studied the morphological change or the molecular biological change in the peritoneum which has very high frequency of implantation sites? According to your lecture, especially for the stromal or epithelial endometrial tissues, there is very little information about the peritoneal site.

Professor Osteen: The pathologist side of me requires that, when I look at a peritoneal lesion, I have to see glands and stroma to diagnose endometriosis. But, the glands and stroma that you see are definitely not normal in their morphology. And, if you assume that a stromal epithelial communication is an important component of progesterone action, then you are not going to have normal stromal epithelial communication. So, what you get in that site is a diminution of the ability of the tissue to respond normally to a steroid signal. But, as you know, it is very variable depending upon the site of the lesion and the degree of inflammatory reactions that are taking place at that site. So that is why I think we have to go beyond that initial steroid signal and come up with another parameter that the cells will respond to. Part of this was driven by the information that a normal luteal phase biopsy that most physicians do, is at a time when the endometrium is under the influence of progesterone. Well, that endometrium doesn't bleed until the end of the cycle. It repairs itself immediately. It repairs itself via turning on and then turning off the MMPs required for that repair process. So, that tells me that there are mechanisms beyond progesterone and steroids which we could use to regulate MMPs to diminish the capacity of this disease to progress. And, that is really what is driving our thought process.

Dr. Kobayashi: My name is Kobayashi and I am a gynecological oncologist. The mechanism of establishment of endometriosis on the tumor invasion and metastasis is very similar. This mechanism has multifactorial steps. Each step may require the action of proteolytic enzymes, one of them is MMP. In the case of tumor cells and associated stroma cells, both are produced not only MMP factor, but also a large amount of TIMPs at the same time. Do you have any data in the case of a normal endometrium or endometrial tissue can produce TIMPs at the same time in in vivo or clinical data?

Professor Osteen: The data we have is some that we have in concert with Kathy Sharpe-Timms at the University of Missouri. I don't know if you know her data, but there is quite a large degree of TIMP-1 expression out in the peritoneal site. But, we also had some insight from in situ hybridization that IL-1 was able to reduce the TIMP expression in focal areas. And so, although again, there may be TIMP that could block MMP action, there are focal abilities to modify the tissue at different sites. Now, whether or not that has something to do with how the tissue implants, we don't know. I will show you some data with the next lecture about blocking MMPs with TIMPs and blocking the establishment of endometriosis in a nude mouse model. We have that data. In the disease process, the degree to which the TIMPs are being expressed or not in different regions is very variable, just like any other protein you look at in these ectopic lesions. It is really kind of a model pattern in which almost every endometrial protein is misexpressed out in the peritoneum. That includes the TIMPs, MMPs, and numerous cytokines. So, part of what we are looking at are certain adhesion molecules that might be involved in implantation that could be involved in the initial attachment sites. So, we have one of our projects looking at that now. And, again, the initial establishment that is like a cancer may or may not be MMP related. There may be other adhesion factors that are more important in early adhesion. So, we are looking at that as well. But, we don't have enough data to speak to that very much right now, even though there are adhesion molecules we think could be involved in that process.

Dr. Kobayashi: Do you think it is also regulated by progesterone?

Professor Osteen: Just like the MMPs, there is an increasing family of TIMPs, and the TIMP-1 appears to be higher by some studies in the late secretory phase prior to menstruation. Other investigators find different expression patterns. So, I don't think we know exactly how the TIMPs are being regulated. TGF-β is a regulator of TIMP. So TGF-β can cause increases in TIMP expression, but out in the peritoneal site, that appears to also be followed by increases in MMPs. So, I think you have the TIMPs and the MMPs quite out of balance in the peritoneal site, which may be one of the confounding issues.

Dr. Kobayashi: May I ask another question? According to your lecture, in the case of coculture of epithelial and stromal cells, progesterone stimulates TGF-β synthesis from stromal cells. And this TGF-β can act on epithelial cells to suppress MMP expression. Is that correct? But, I have read in your papers, in the case of epithelial cells alone, TGF-β inhibits MMP expression while progesterone does not inhibit MMP expression. Is that correct?

Professor Osteen: In isolated epithelial cells, you can treat isolated epithelial cells with progesterone and you can't suppress MMP-7, but isolated epithelial cells can be treated with TGF-β and you will suppress, in human endometrial epithelial or keratinocytes. So, the MMP-7 appears to be highly regulated in epithelial cells by TGF-β. In co-cultures, treatment with progesterone in a co-culture allows there to be a suppression of MMP-7, and blocking TGF-β blocks that effect. However, the TGF-β production can also be produced by epithelial cells in certain conditions. And so, there appears to be a dynamic equilibrium that might involve progestin actions, and might involve retinoic acid as well, in order to change the way the cells respond to TGF-β. For example, TGF-β actually stimulates fibroblast proliferation and MMP expression in wound repair. So, TGF-β can be a stimulatory agent as well as an inhibitory agent for MMPs depending upon the state of differentiation of the cells. So, that is one of the reasons why I think many investigators have shown different results with TGF-β. The differentiated states of the cell can determine the response of the cell to TGF-β. For example, prior treatments of certain fibroblasts with retinoic acid changes the subsequent response to TGF-β. So, the stages of the cell's differentiation can have a lot to do with the response of that cell to TGF-β, including regulation of MMPs. And it is poorly understood exactly what factors are involved in that interaction.

Professor Terakawa: So Professor Osteen, do you propose TGF-β is a progesterone-induced stromal factor?

Professor Osteen: TGF-β is a progesterone-induced endometrial factor that is produced apparently by both stroma and epithelial cells in different situations. The differentiation of the stroma to the decidua increases decidual TGF-β_2 specifically.

Dr. Momoeda: Thank you for your excellent presentation. I am still interested in the retinoic acid system and the regulation of MMPs. Considering, however, the prophylactic medical therapy, the retinoic acid or vitamin A is a candidate of some medicine for the therapy of endometriosis. The differential expression of the retinoic acid from epithelial cells and the stroma cell is completely different in the menstrual cycle as shown in your figures. Do you have any data or information the differential regulation of the retinoic acid system?

Professor Osteen: No, in a word. What we have found and what we have mostly published has been in the rat system, and I don't know if you have seen any of that work with David Ong's group at Vanderbilt University.

Dr. Momoeda: But, I read your paper on the retinoic acid binding proteins in the endometrium.

Professor Osteen: Yes, and what we find is that in the rat system and in the human system, which are the only two where people have been looking at the retinoic acid synthesis, I believe, is that the epithelial cell has the capability to synthesize retinoic acid from retinol precursors. The stromal cell does not have that capacity, but gains that capacity upon decidualization, either during in vivo decidualization as we have shown in the rat, or during in vitro decidualization which we have recently shown in the human. So, this is another example of the decidual cell in many ways becoming epithelioid in terms of what all the cell is able to do as a decidua. And so, epithelial cells may synthesize retinoic acid as a component of normal cell cycle or turnover that occurs in the endometrium at the end of the proliferative phase. And there are indications that the MMPs are suppressed quite a bit by some factors that are not progesterone dependent during the late proliferative phase. But, as the stromal cell begins to decidualize, there is a huge increase in the capacity of a decidual cell to synthesize retinoic acid. So, progesterone is driving that reaction via some mechanism, but what MMP enzyme system might be affected we don't know yet.

Dr. Momoeda: So you don't have data on the producing cell for retinoic acid in the stroma. Which cell is producing the retinoic acid in the stroma in the elementary secretory phase?

Professor Osteen: It is the differentiated stromal cell, yes. That is, isolated stromal cells decidualized in vitro develop that capacity.

Dr. Kurachi: You mentioned that the progesterone regulation of TGF-β depends on the stage of the cells, right? Does that mean the regulation is indirect? What is the mechanism? What is the molecular mechanism of progesterone regulation of TGF-β expression?

Professor Osteen: It is a great question, and it is a question that applies equally to the MMPs. And several of the MMPs that are suppressed by progesterone do not have progesterone-response elements in the promotor of the gene. TGF-β_2 does not have a progesterone-response elements in the gene. We have data, preliminary data, that there may be an indirect nontraditional mechanism. We have transfected some cells with the TGF-β_2 promoter and were able to induce promoter activity with the combination of progesterone and cyclic AMP. And that is probably some of the first data that shows progesterone might be regulating genes that don't possess a true PRE or even a half site. And so, there is a sort of emerging field among some people in endocrinology that progesterone and estrogen may be activating and regulating genes through the regulation of co-activators and co-repressors through mechanisms that we don't yet quite understand. And that would be the only answer to your question. TGF-β_2 appears to be a differentiation factor whose expression is driven by progesterone, but how direct the effect of progesterone is on the gene, we don't know yet.

Session 1
Review

Gynecologic and Obstetric Investigation

Gynecol Obstet Invest 1999;48(suppl 1):14–20

Angiogenesis in Endometriosis and Angiogenic Factors

Jiro Fujimoto Hideki Sakaguchi Reiko Hirose Hongwu Wen Teruhiko Tamaya

Department of Obstetrics and Gynecology, Gifu University School of Medicine, Gifu, Japan

Key Words
Angiogenesis · Basic FGF · VEGF · PD-ECGF · Ovarian endometriosis · Gynecological cancers

Abstract
Among angiogenic factors, VEGF secreted from activated macrophages under the influence of ovarian steroids, IL-8 expressed in endometrial stromal cells, and basic FGF expressed in endometriotic tissue and PD-ECGF expressed in lining epithelial cells independently of the sex steroidal milieu might contribute to the characteristic advancement of angiogenic lesions in endometriosis in individual manners.

Introduction

Newly developed capillary network formation from the original vessel is designated as angiogenesis. Generally, turnover of capillary endothelial cells is extremely slow, in the order of months or years in physiological angiogenesis, while the turnover in ovary and uterine endometrium among normal tissues, and in tumors and tumor-like tissues is altered to a rapid state. Angiogenesis is essential to the advancement of pelvic endometriosis [1–5] and solid tumors greater than 2 mm in diameter [6]. Angiogenesis consists of the following steps: dissolution of basement membrane by proteases derived from vascular endothelial cells activated by angiogenic factors, migration and proliferation of the endothelial cells, and capillary tube formation [7] (fig. 1). Each step is regulated by various angiogenic factors (tab. 1).

Some angiogenic factors that have been identified are shown in table 1. PD-ECGF was cloned as a novel angiogenic factor (45 kDa polypeptide) from human platelet [8]. Thereafter, PD-ECGF was completely identified with thymidine phosphorylase (TP) [9, 10]. PD-ECGF/TP does not stimulate the growth of endothelial cells, but rather, chemotaxis of them, and induces angiogenesis in vivo with the activation of TP as an enzyme [11, 12]. The functions of angiogenic factors were studied in tumors initially and later in endometriosis. Thus we describe the functions in tumors first, and then those in endometriosis.

Angiogenesis in Gynecologic Tumors

Various tumor-derived angiogenic factors stimulate tumor growth and secondary spreading. It has been demonstrated that basic fibroblast growth factor (FGF), vascular endothelial growth factor (VEGF), and PD-ECGF mainly contribute to growth in tumors. A novel angiogenic factor, PD-ECGF, is expressed in lymph nodes, peripheral lymphocytes, spleen, lung, liver, placenta [13], and uterine endometrium [14, 15] among normal tissues. Among sol-

KARGER
Fax +41 61 306 12 34
E-Mail karger@karger.ch
www.karger.com

0378–7346/99/0485–0014$17.50/0

Accessible online at:
www.karger.com/journals/goi

Jiro Fujimoto, MD, PhD
Department of Obstetrics and Gynecology
Gifu University School of Medicine
40 Tsukasa-machi, Gifu City 500-8705 (Japan)
Tel. +81 58 267 2631, Fax +81 58 265 9006

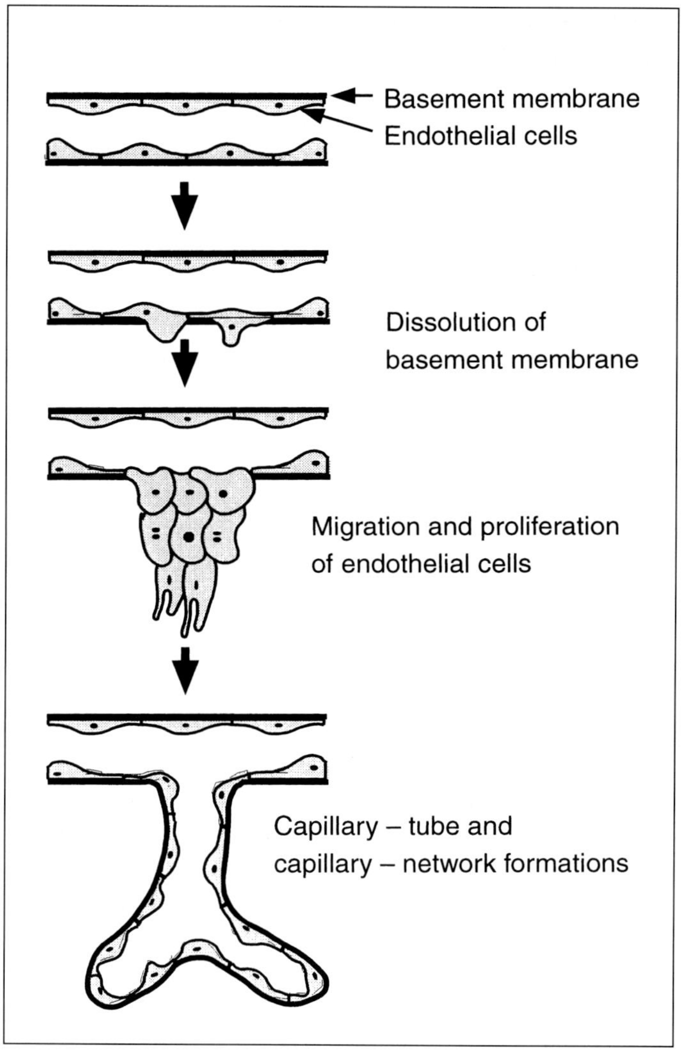

Fig. 1. The process of angiogenesis.

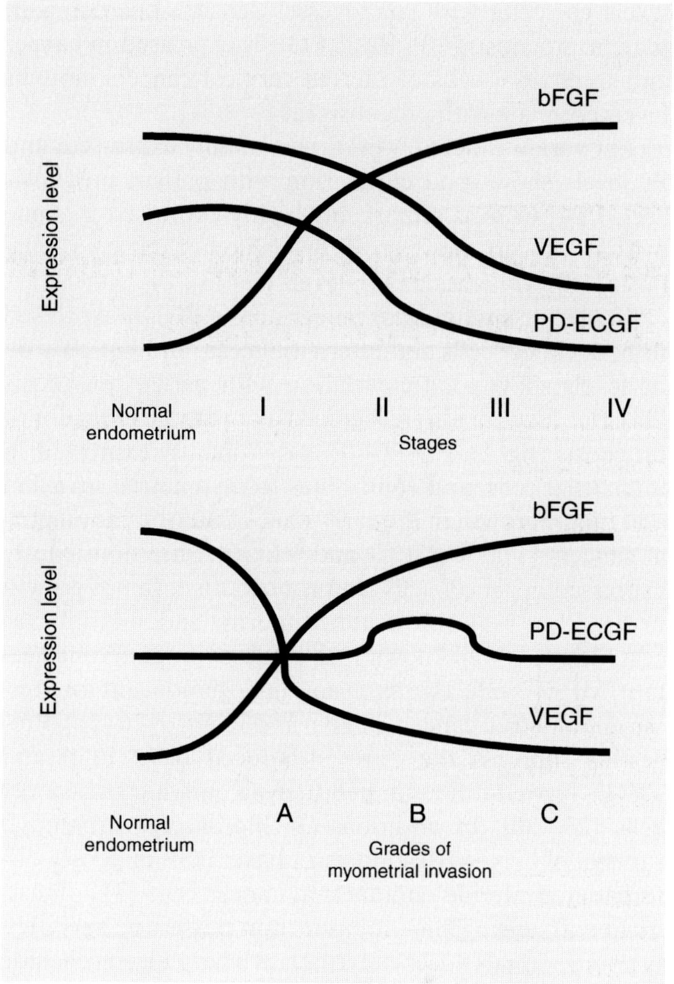

Fig. 2. Angiogenic factors in uterine endometrial cancers. A = Carcinoma in endometrium; B = invasion of ≤ 1/2 the depth of myometrium; C = invasion of > 1/2 the depth of myometrium.

id tumors, PD-ECGF is expressed in malignant gliomas, thyroid tumors, cancers of the breast, esophagus, stomach, colon, pancreas, gall bladder, kidney, bladder, lung [13], uterine cervix [16] and endometrium [17], and ovary [18].

In uterine cervical cancers, PD-ECGF is recognized as a reliable prognostic factor [16]. PD-ECGF is dominantly expressed in interstitial cells, and its levels correlate with microvessel density in squamous cell carcinomas of the uterine cervix, in which 5′-deoxy-5-fluorouridine might be highly effective [16]. PD-ECGF expressed in metastatic lymph node lesions of uterine cervical cancers is a prognostic indicator [19]. VEGF isoforms $VEGF_{165}$ and $VEGF_{121}$ are mainly expressed in cancer cells of uterine cervical cancers, especially adenocarcinomas, and their

Table 1. Angiogenic factors

Basic fibroblast growth factor
Vascular endothelial growth factor
Placenta growth factor
Epidermal growth factor
Transforming growth factor-α and -β
Platelet-derived growth factor
Platelet-derived endothelial cell growth factor
Hepatocyte growth factor
Tumor necrosis factor-α
Pleiotropin
Proliferin
Angiogenin
Estradiol
Interleukin-8

levels correlate with microvessel density, but not with patient prognosis [20]. Basic FGF is expressed in cancer and interstitial cells of uterine cervical cancers, and its levels correlate with prognosis [21].

In ovarian cancers, $VEGF_{165}$ is mainly expressed, and its levels show good correlation with patient prognosis [22]. PD-ECGF is remarkably highly expressed in some ovarian cancers, however its levels do not relate to prognosis [18], while basic FGF levels do [23].

In uterine endometrial cancers, basic FGF is expressed in both cancer cells and interstitial cells, and not only do its levels show good correlation with patient prognosis [24], but they are also recognized as a reliable indicator of prognosis (fig. 2). PD-ECGF is dominantly expressed in interstitial cells, and contributes to myometrial invasion and tumor growth in the early stage of uterine endometrial cancers [17]. $VEGF_{165}$ and $VEGF_{121}$ are dominantly expressed in cancer cells, and contribute to tumor growth in stage I of well-differentiated uterine endometrial cancers [25]. It is well known that medroxyprogesterone acetate (MPA) is effective as an antiangiogenic agent for uterine endometrial cancers in vivo and in vitro [26–28]. Progestins suppress the estrogen-induced basic FGF and VEGF in well-differentiated uterine endometrial cancer cells [29, 30]. In addition, *O*-(chloroacetyl-carbamoyl) fumagillol (TNP470) suppresses basic FGF in poorly differentiated uterine endometrial cancer cells [31]. Treatment with angiogenic inhibitors that possess no specificity to individual cancer cells (such as TNP470, ginsenoside Rb2, toremifene, ICI 182,780, etc.) should be recommended for patient care.

Angiogenesis in Endometriosis

Basic FGF [32–34], VEGF [3, 35, 36], interleukin (IL)-8 [2, 37–39] and PD-ECGF [40] have been recognized as positive factors in the advancement of endometriosis as follows: Basic FGF and its mRNA, and basic FGF receptor mRNA, are detected in endometriotic tissue [32, 33], and basic FGF levels correlate with endometrioma diameter [33]. Conversely, the concentration of basic FGF in peritoneal fluid of women with endometriosis is so low as to be unable to bind to its receptor [34]. Therefore, basic FGF might promote endometriosis advancement in endometriotic tissue in autocrine and paracrine manners.

Peritoneal fluid concentrations of VEGF are higher in women with moderate to severe endometriosis than in women with minimal to mild endometriosis [3]. Activated macrophages are a major source of VEGF in endometriosis, and this expression is regulated directly by ovarian steroids [35]. Cyclic variations in VEGF levels are seen in peritoneal fluid of patients with endometriosis [36]. Therefore, elevated VEGF secreted from activated macrophages into peritoneal fluid might contribute to vascular pathogenesis of endometriosis.

IL-8 concentrations are high in the peritoneal fluid of women with endometriosis [2, 38], and contribute to the pathogenesis of endometriosis by promoting the neovascularization of ectopic endometrial implants [2]. Cultured mesothelial cells express cytokine-inducible IL-8 mRNA and secreted IL-8 [37]. IL-8 induces proliferation [37], and furthermore reverses the proliferation-inhibitory effect of IL-8 antisense oligonucleotides in uterine endometrial stromal cells [39]. Therefore, IL-8 might act as an autocrine growth factor in the angiogenic pathogenesis of endometriosis.

PD-ECGF was expressed in lining epithelial cells of ovarian endometriomas and in interstitial cells of the subepithelial area [40]. Consistent with subepithelial PD-ECGF expression, factor VIII-related antigen was expressed in vessel cells of the subepithelial area, which might be recognized as endometriomal angiogenesis [40]. In general, the response to estrogen and progesterone is partially altered in endometriosis [41, 42]. The expression of PD-ECGF persisted in endometriotic endometrium during the menstrual cycle [15, 40]. Therefore, PD-ECGF might contribute to the growth of ovarian endometriomas via subepithelial angiogenesis independently of the sex steroidal milieu.

Conclusion

Elevated levels of VEGF secreted from activated macrophages to peritoneal fluid may contribute to vascular pathogenesis of endometriosis under the influence of ovarian steroids. Conversely, basic FGF might promote endometriosis advancement in endometriotic tissue in autocrine and paracrine manners independently of the sex steroidal milieu. IL-8 in endometrial stromal cells is likely to act as an autocrine growth factor in the angiogenic pathogenesis of endometriosis. PD-ECGF was persistently expressed in interstitial cells of the subepithelial area of ovarian endometriomas with angiogenesis and lack of sex steroidal regulation. The angiogenic factors VEGF, basic FGF, IL-8, and PD-ECGF might contribute to the characteristic advancement of angiogenic lesions in endometriosis in an individual fashion.

References

1 Oosterlynck DJ, Meuleman C, Sobis H, Vandeputte M, Koninckx PR: Angiogenic activity of peritoneal fluid from women with endometriosis. Fertil Steril 1993;59:778–782.
2 Ryan IP, Tseng JF, Schriock ED, Khorram O, Landers DV, Taylor RN: Interleukin-8 concentrations are elevated in peritoneal fluid of women with endometriosis. Fertil Steril 1995;63:929–932.
3 Shifren JL, Tseng JF, Zaloudek CJ, Ryan IP, Meng YG, Ferrara N, Jaffe RB, Taylor RN: Ovarian steroid regulation of vascular endothelial growth factor in the human endometrium: implications for angiogenesis during the menstrual cycle and in the pathogenesis of endometriosis. J Clin Endocrinol Metab 1996;81:3112–3118.
4 Smith SK: Angiogenesis. Semin Reprod Endocrinol 1997;15:221–227.
5 Taylor RN, Ryan IP, Moore ES, Hornung D, Shifren JL, Tseng JF: Angiogenesis and macrophage activation in endometriosis. Ann NY Acad Sci 1997;828:194–207.
6 Folkman J: Tumor angiogenesis. Adv Cancer Res 1985;43:175–203.
7 Folkman J, Haudenschild C: Angiogenesis in vitro. Nature 1980;288:551–556.
8 Ishikawa F, Miyazono K, Hellman U, Drexler H, Wernstedt C, Hagiwara K, Usuki K, Takaku F, Risau W, Heldin CH: Identification of angiogenic activity and the cloning and expression of platelet-derived endothelial cell growth factor. Nature 1989;338:557–562.
9 Furukawa T, Yoshimura A, Sumizawa T, Haraguchi M, Akiyama S: Angiogenic factor. Nature 1992;356:668.
10 Sumizawa T, Furukawa T, Haraguchi M, Yoshimura A, Takeyasu A, Ishizawa M, Yamada Y, Akiyama S: Thymidine phosphorylase activity associated with platelet-derived endothelial cell growth factor. J Biochem 1993;114:9–14.
11 Haraguchi M, Miyadera K, Uemura K, Sumizawa T, Furukawa T, Yamada K, Akiyama S, Yamada Y: Angiogenic activity of enzymes. Nature 1994;368:198.
12 Miyadera K, Sumizawa T, Haraguchi M, Yoshida H, Konstanty W, Yamada Y, Akiyama S: Role of thymidine phosphorylase activity in the angiogenic effect of platelet-derived endothelial cell growth factor/thymidine phosphorylase. Cancer Res 1995;55:1687–1690.
13 Yoshimura A, Kuwazuru Y, Furukawa T, Yoshida H, Yamada K, Akiyama S: Purification and tissue distribution of human thymidine phosphorylase; high expression in lymphocyte, reticulocytes and tumors. Biochem Biophys Acta 1990;1034:107–113.
14 Osuga Y, Toyoshima H, Mitsuhashi N, Taketani Y: The presence of platelet-derived endothelial cell growth factor in human endometrium and its characteristic expression during the menstrual cycle and early gestational period. Hum Reprod 1995;10:989–993.
15 Fujimoto J, Ichigo S, Sakaguchi H, Hirose R, Tamaya T: Expression of platelet-derived endothelial cell growth factor (PD-ECGF) and its mRNA in uterine endometrium during the menstrual cycle. Mol Hum Reprod 1998;4:509–513.
16 Fujimoto J, Sakaguchi H, Ichigo S, Hirose R, Tamaya T: Expression of platelet-derived endothelial cell growth factor (PD-ECGF) and its mRNA in uterine cervical cancers. Br J Cancer 1999;79:1249–1254.
17 Fujimoto J, Ichigo S, Sakaguchi H, Hirose R, Tamaya T: Expression of platelet-derived endothelial cell growth factor (PD-ECGF) and its mRNA in uterine endometrial cancers. Cancer Lett 1998;130:115–120.
18 Fujimoto J, Ichigo S, Sakaguchi H, Hirose R, Tamaya T: Expression of platelet-derived endothelial cell growth factor (PD-ECGF) and its mRNA in ovarian cancers. Cancer Lett 1998;126:83–88.
19 Fujimoto J, Sakaguchi H, Hirose R, Wen H, Tamaya T: Clinical implication of expression of platelet-derived endothelial cell growth factor (PD-ECGF) in metastatic lesions of uterine cervical cancers. Cancer Res 1999;54:3041–3044.
20 Fujimoto J, Sakaguchi H, Hirose R, Ichigo S, Tamaya T: Expression of vascular endothelial growth factor (VEGF) and its mRNA in uterine cervical cancers. Br J Cancer 1999;80:827–833.
21 Fujimoto J, Ichigo S, Hori M, Hirose R, Sakaguchi H, Tamaya T: Expression of basic fibroblast growth factor and its mRNA in advanced uterine cervical cancers. Cancer Lett 1997;111:21–26.
22 Fujimoto J, Sakaguchi H, Hirose R, Ichigo S, Tamaya T: Biological implication of the expression of vascular endothelial growth factor (VEGF) subtypes in ovarian cancers. Cancer 1998;83:2527–2533.
23 Fujimoto J, Ichigo S, Hori M, Hirose R, Sakaguchi H, Tamaya T: Expression of basic fibroblast growth factor and its mRNA in advanced ovarian cancers. Eur J Gynaecol Oncol 1997;18:349–352.
24 Fujimoto J, Hori M, Ichigo S, Tamaya T: Expressions of basic fibroblast growth factor and its mRNA in uterine endometrial cancers. Invas Metast 1995;15:203–210.
25 Fujimoto J, Ichigo S, Hirose R, Sakaguchi H, Tamaya T: Expression of vascular endothelial growth factor (VEGF) and its mRNA in uterine endometrial cancers. Cancer Lett 1998;134:15–22.
26 Fujimoto J, Fujita H, Hosoda S, Okada H, Tanaka NG: Effect of medroxyprogesterone acetate on secondary spreading of endometrial cancer. Invas Metast 1989;9:209–215.
27 Fujimoto J, Hosoda S, Fujita H, Okada H: Inhibition of tumor angiogenesis activity by medroxyprogesterone acetate in gynecologic malignant tumors. Invas Metast 1989;9:269–277.
28 Jikihara H, Terada N, Yamamoto R, Nishikawa Y, Tanizawa O, Matsumoto K, Terakawa N: Inhibitory effect of medroxyprogesterone acetate on angiogenesis induced by human endometrial cancer. Am J Obstet Gynecol 1992;167:207–211.
29 Fujimoto J, Hori M, Ichigo S, Hirose R, Tamaya T: Antiestrogenic compounds inhibit estrogen-induced expression of basic fibroblast growth factor and its mRNA in well differentiated endometrial cancer cells. Gen Pharmacol 1997;28:215–219.
30 Fujimoto J, Sakaguchi H, Hirose R, Ichigo S, Tamaya T: Progestins suppress estrogen-induced expression of vascular endothelial growth factor (VEGF) subtypes in uterine endometrial cancer cells. Cancer Lett 1998;134:15–22.
31 Fujimoto J, Hori M, Ichigo S, Hirose R, Sakaguchi H, Tamaya T: Plausible novel therapeutic strategy of uterine endometrial cancer with reduction of basic fibroblast growth factor secretion by progestin and O-(chloroacetyl-carbamoyl) fumagillol [TNP-470; AGM-1470]. Cancer Lett 1997;113:187–194.
32 Ferriani RA, Charnock-Jones DS, Prentice A, Thomas EJ, Smith SK: Immunohistochemical localization of acid and basic fibroblast growth factors in normal human endometrium and endometriosis and the detection of their mRNA by polymerase chain reaction. Hum Reprod 1993;8:11–16.
33 Di Blasio AM, Centinaio G, Carniti C, Somigliana E, Vigano P, Vignali M: Basic fibroblast growth factor messenger ribonucleic acid levels in eutopic human endometrial stromal cells as assessed by competitive polymerase chain reaction amplification. Mol Cell Endocrinol 1995;115:169–175.
34 Huang JC, Papasakelariou C, Dawood MY: Epidermal growth factor and basic fibroblast growth factor in peritoneal fluid of women with endometriosis. Fertil Steril 1996;65:931–934.
35 McLaren J, Prentice A, Charnock-Jones DS, Millican SA, Muller KH, Sharkey AM, Smith SK: Vascular endothelial growth factor is produced by peritoneal fluid macrophages in endometriosis and is regulated by ovarian steroids. J Clin Invest 1996;98:482–489.
36 McLaren J, Prentice A, Charnock-Jones DS, Smith SK: Vascular endothelial growth factor (VEGF) concentrations are elevated in peritoneal fluid of women with endometriosis. Hum Reprod 1996;11:220–223.

37 Iwabe T, Harada T, Tsudo T, Tanikawa M, Onohara Y, Terakawa N: Pathogenic significance of increased levels of interleukin-8 in the peritoneal fluid of patients with endometriosis. Fertil Steril 1998;69:924–930.
38 Arici A, Tazeke SI, Attar E, Kliman HJ, Olive DL: Interleukin-8 concentration in peritoneal fluid of patients with endometriosis and modulation of interleukin-8 expression in human mesothelial cells. Mol Hum Reprod 1996;2:40–45.
39 Arici A, Seli E, Zeyneloglu HB, Senturk LM, Oral E, Olive D:Interleukin-8 induces proliferation of endometrial stromal cells: a potential autocrine growth factor. J Clin Endocrinol Metab 1998;83:1201–1205.
40 Fujimoto J, Sakaguchi H, Hirose R, Tamaya T: Expression of platelet-derived endothelial cell growth factor (PD-ECGF) in ovarian endometriosis. J Clin Endocrinol Metab 1999;84:359–362.
41 Tamaya T, Motoyama T, Ohno Y, Ide N, Tsurusaki T, Okada H: Steroid receptor levels and histology of endometriosis and adenomyosis. Fertil Steril 1979;31:396–400.
42 Fujimoto J, Ichigo S, Hirose R, Sakaguchi H, Tamaya T: Expression of estrogen receptor wild type and exon 5 splicing variant mRNAs in normal and endometriotic endometria during the menstrual cycle. Gynecol Endocrinol 1997;11:11–16.

Post-Presentation Discussion

Professor Terakawa: Thank you very much for a very nice presentation. The paper is open for discussion. Questions and comments?

Professor Osteen: Given the lack of the cyclic variability in the menstrual cycle and the changes that were expressed in the endometrioma, what is your view of the relationship between these steroids and the growth, versus other factors?

Dr. Fujimoto: Of course, we believe that estrogen stimulation is very important for the development of endometriosis. However, endometriosis has some lack of steroidal regulation. The molecular mechanisms of this lack of steroidal regulation are as follows: Wild-type estrogen and progesterone receptors are relatively lower in endometriosis. Furthermore, many kinds of progesterone receptor splicing variants express as a dominant negative in endometriosis, which lead to some damage of the inhibitory effect on estrogen stimulation. Of course, the estrogen stimulation is essential to grow endometriosis, but I think, at least, PD-ECGF contribution for endometriosis is not correlated with steroidal regulation.

Professor Terakawa: Does estradiol suppress PD-ECGF expression? Is that what you mean?

Dr. Fujimoto: No, I don't think so. Today I showed the reverse expression of PD-ECGF in uterine endometrium with estradiol in serum during the menstrual cycle. However, the expression of PD-ECGF in well-differentiated endometrial cancer cells, Ishikawa cell line, is increased by estrogen and progesterone. Dr. Osuga, with Professor Taketani, also already published the remarkable induction of PD-ECGF expression at the early gestation. Therefore, I can conclude that estrogen stimulation must be required to get the high expression of PD-ECGF by progesterone. Additionally, I don't know whether it is unfortunately or fortunately that MPA does not stimulate PD-ECGF expression.

Professor Terakawa: One more question, what about expression of mRNA and protein of PD-ECGF in eutopic and ectopic endometrium? Is there a big difference there?

Dr. Fujimoto: The level of PD-ECGF mRNA is biologically correlated with the level of PD- ECGF with the time lag to accumulate the protein PD-ECGF in uterine endometrium. As I did not get any variant of PD-ECGF in eutopic and ectopic endometrium so far, both levels must be correlated.

Professor Terakawa: My question is the level. Was there a difference in expression level in the eutopic and ectopic?

Dr. Fujimoto: I see. They are similar levels.

Dr. Kobayashi: I'm sorry, I am not familiar with this unusual factor. But, usually by immunohistochemistry, epithelial cells and the stromal cells also produce these factors. Is there any correlation between the severity of the disease, endometriosis, and the change of immunohistochemical staining?

Dr. Fujimoto: So far, I didn't get any chages of PD-ECGF immunohistochemical staining pattern with the grade of endometriosis. Because this enzyme works as a dimer and also makes many metabolites, and it is hard to analyze the correlation between the stage of endometriosis and only the PD-ECGF immunohistochemical staining pattern precisely right now.

Dr. Kobayashi: And do you think the interstitial cell corresponds to the macrophage?

Dr. Fujimoto: I hope so. I will study that.

Dr. Kobayashi: And the production of this androgen factor for macrophage is also regulated by steroids or such growth factor?

Dr. Fujimoto: I think so.

Dr. Kobayashi: In the case of VEGF, it does significantly stimulate the production of, for example, cancer cells, or several interstitial cells. But this androgen factor, it does not stimulate the LH or HCG?

Dr. Fujimoto: I don't know.

Dr. Kobayashi: Maybe produced by progesterone, because of the secretory phase.

Dr. Fujimoto: My answer is yes. I am checking the expression of PD-ECGF and IL-8 in uterine cervical cancers. IL-8 is dominantly excreted from macrophages in uterine cervical cancers. I am not sure of the dominant source of PD-ECGF and VEGF in each disease. I will be able to address this point in about six months.

Dr. Kobayashi: In general, in the case of cancers associated with inflammatory cells including macrophages or neutrophils, do such patients have a good prognosis?

Dr. Fujimoto: VEGF is not recognized as a prognostic indicator associated with inflammatory cells, but IL-8 is and, additionally, PD-ECGF probably is. However, please wait for the answer right now, because we are studying the angiogenesis derived from macrophage.

Dr. Suginami: You talked about a large number of growth factors. I am rather confused. But this time you talked about growth factors and in the other presentation you also talked about other growth factors in endometriosis. Which is the most important growth factor? What is your opinion?

Dr. Fujimoto: There is four major angiogenic factors, basic FGF, VEGF, IL-8 and PD-ECGF, and HGF in brackets in endometriosis. From the aspect of steriodal influence, some facts about angiogenic factors are as follows: The basic FGF according to well-known articles does not receive the impact of estrogen or progesterone in uterine endometrium. However, personally, according to my data, basic FGF is persistently regulated and impacted by those two sex steroids in uterine endometrium. Progesterone transiently down-regulates the estrogen-induced VEGF in well-differentiated endometrial cancer Ishikawa cells. Professor Terakawa's lab has researched IL-8 contribution in endometriosis, which might be also subtle. Therefore, I think basic FGF does have a basic contribution to angiogenesis and VEGF might contribute to some transient modification of angiogenesis in endometriosis under the steroidal influence, and IL-8 and PD-ECGF might also contribute to some persistent modifications of angiogenesis independently upon steroidal influence.

Dr. Suginami: In the cycle where the PD-ECGF messenger RNA level was discussed: the level of the products was discussed, in the messenger RNA at the time of ovulation it would go down during the retaining phase, it was increasing. I wasn't able to follow your message there.

Dr. Fujimoto: There is the time lag in the alteration of PD-ECGF and its mRNA to accumulate the protein in uterine endometrium. Therefore, in biological terms, if you look at the protein level, I think it should reflect the function.

Dr. Suginami: So, what you are trying to say is that at the ovulation time the messenger RNA expression level is down. It is down-regulated, and where is that timing in the product expression cycle?

Dr. Fujimoto: The mRNA transiently goes down toward ovulation, and then the protein slightly goes down after ovulation.

Dr. Suginami: Oh, does it go down?

Dr. Fujimoto: Yes, you could see it. However, the review did not really mention much about this aspect. The mRNA level does not alter during the proliferative phase, transiently decreases toward ovulation, and does not really alter during the secretory phase. However the protein level does not alter during the proliferative phase, transiently slightly decreases after ovulation, and remarkably increases during the secretory phase.

Dr. Suginami: The messenger RNA level goes down like this, and the protein level, it rises?

Dr. Fujimoto: No, that is not what I mean to say. What I wanted to say is that the product from the mRNA accumulates and rises.

Dr. Suginami: You are talking about the secretory phase?

Dr. Fujimoto: Yes, I am. The mRNA level shows like a 'V', namely the level during the proliferative and secretory phases are similar. However, the protein level alters in a different way with the accumulation of synthesized protein.

Dr. Suginami: How can you explain this difference?

Dr. Fujimoto: Even if the mRNA level persists, the protein will continue to accumulate during the secretory phase. Therefore, it will continue to rise in this matter, and then menstruation starts and it goes back to the beginning of the cycle.

Dr. Kubota: Let me digress slightly. When we look at the stages of endometriosis, we use the tumor markers, and you discussed the PD-ECGF expression and the tumor markers. Have you studied the relationship between this factor and the tumor markers?

Dr. Fujimoto: That is a very important question in the PD-ECGF expression recognized as a tumor marker. In my research, in fact, what you have asked is the bottleneck or the barrier. We have only studied the expression of PD-ECGF in ovarian endometrioma among various endometriosis except for some tumors. We have not studied the correlation of the expression with the grade or the stage of endometriosis. We have also studied the expression of PD-ECGF in peritoneal fluid, but we could not find a good correlation with the stage. In the case of tumors, especially ovarian cancers, if the capsule is already disrupted, high PD-ECGF is detected. But another question is whether that is functional or not. If you look at the serum concentration of PD-ECGF, some information must be detected in uterine cervical cancers. However, if they are ovarian cancers, we don't expect high correlation with the stage or patient prognosis.

Professor Terakawa: Thank you, Dr. Fujimoto.

Session 1
Review

Gynecologic and Obstetric Investigation

Gynecol Obstet Invest 1999;48(suppl 1):21–28

Expression of Aromatase Cytochrome P450 in Eutopic Endometrium and Its Application as a Diagnostic Test for Endometriosis

Jo Kitawaki Izumi Kusuki Hisato Koshiba Katsumi Tsukamoto Hideo Honjo

Department of Obstetrics and Gynecology, Kyoto Prefectural University of Medicine, Kyoto, Japan

Key Words
Endometriosis · Infertility · Aromatase cytochrome P450 · Estrogen · Endometrial biopsy · Diagnostic test · Immunohistochemistry

Abstract
Endometriotic implants, like other estrogen-dependent tumors, contain both estrogen receptors and aromatase cytochrome P450 (P450arom), suggesting that at a local level, endometriotic implants produce estrogens, which may be involved in tissue growth through interaction with the estrogen receptors. P450arom is also expressed in the eutopic endometria of patients with endometriosis, adenomyosis, and/or leiomyomas, whereas neither P450arom protein nor mRNA is expressed in the eutopic endometria of normal menstruating women with cervical carcinoma in situ yet showing no other gynecological disease (disease-free). Examination of P450arom expression in endometrial biopsy specimens enables the physician to discriminate between the presence and absence of endometriosis, and may be used as an initial screening at outpatient infertility clinics.

Introduction

Endometriosis develops mostly in women of reproductive age and regresses after menopause. Occurrence of endometriosis before menarche has not been reported. Considering the regression of lesions only, ovariectomy is the most effective therapy among various surgical and endocrine therapies. Suppression of estrogen level derived by danazol or gonadotropin-releasing hormone agonists provides the regression of lesions, however, recovery of the estrogen level after discontinuation of the therapies causes the relapse of lesions. Endometriosis may relapse in postmenopausal women who have been treated with the estrogen replacement therapy. This evidence undoubtedly suggests that endometriosis grows and regresses in an estrogen-dependent fashion.

Estrogen-dependent tumor tissues mostly contain estrogen, progesterone, and androgen receptors as shown by hormone-ligand binding assays [1] and immunohistochemistry [2]. Interestingly, estrogen-dependent diseases of the uterus such as endometrial cancer [3–5], leiomyomas [6, 7], and adenomyosis [8, 9] also contain aromatase, an enzyme that catalyzes conversion of androgens to estrogens, suggesting that the local estrogen production may increase the estrogen concentration, which further

KARGER

Fax + 41 61 306 12 34
E-Mail karger@karger.ch
www.karger.com

0378–7346/99/0485–0021$17.50/0

Accessible online at:
www.karger.com/journals/goi

Jo Kitawaki, MD, PhD
Department of Obstetrics and Gynecology
Kyoto Prefectural University of Medicine
465 Kajii-cho, Kamigyo-ku, Kyoto 602-8566 (Japan)
Tel. +81 75 251 5560, Fax +81 75 212 1265, E-Mail kitawaki@koto.kpu-m.ac.jp

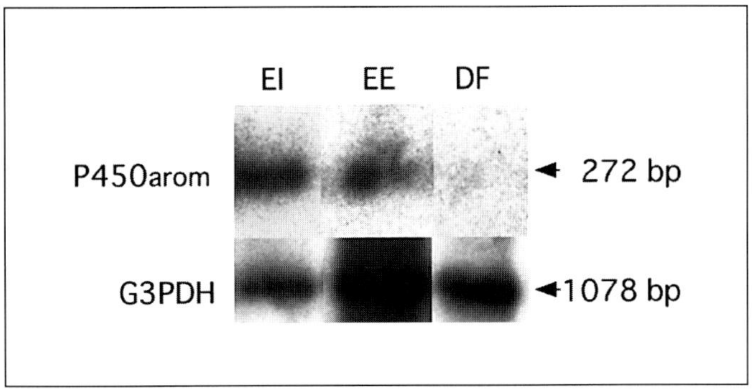

Fig. 1. Southern blots of aromatase cytochrome P450 (P450arom) transcripts after reverse transcriptase polymerase chain reaction amplification. Total RNA samples obtained from endometriotic tissue in a peritoneal vesicle implant (EI), eutopic endometrium obtained from a patient with endometriosis (EE), and eutopic endometrium from a patient with cervical carcinoma in situ but without any other gynecological disease (disease-free) (DF) were reverse transcribed, amplified with P450arom or G3PDH primers, and hybridized with the ^{32}P-labeled corresponding probes.

simulates the growth of these tissues. We review the role of aromatase in endometriotic implants and the eutopic endometrium.

Expression of P450arom in Endometriotic Implants

Because of very low levels of aromatase activity and small tissue volume, it has been difficult to detect aromatase activity in endometriotic implants. Reverse transcriptase polymerase chain reaction (RT-PCR) and immunohistochemical techniques have enabled the detection of the mRNA and protein of aromatase cytochrome P450 (P450arom), the major component of aromatase, in such tissues. Using RT-PCR, Noble et al. [10] have detected P450arom transcripts in peritoneal endometriotic implants and have demonstrated that the endometriotic implants use both the adipose-type promoter I.4 and gonadal-type promoter II for P450arom expression. We have detected P450arom transcripts in endometriotic tissue in a peritoneal vesicle implant (fig. 1) [11]. We have then studied the immunohistochemical localization of P450arom in endometriotic and adenomyotic tissues by using a specific antibody and have shown that P450arom is localized exclusively in the cytoplasm of glandular cells (fig. 2A) [11]. Although adenomyosis is of a separate entity, P450arom transcripts are also detected in adenomyotic tissue homogenates, and P450arom protein is immunolocalized in the glandular cells of adenomyotic tissues [11]. The aromatase activity in adenomyotic tissue homogenates is inhibited by the addition of a specific monoclonal antibody to P450arom and by addition of some aromatase inhibitors, including danazol. This suggests that the enzymatic property of aromatase in adenomyotic tissue is similar to that seen in other human tissues [11]. These findings suggest that at a local level, endometriotic implants produce estrogens, which may be involved in tissue growth through interaction with the estrogen receptors.

P450arom Is Expressed in the Eutopic Endometria of Patients with Endometriosis, but Not in Those of Disease-Free Women

There has been evidence that the eutopic endometrium of patients with endometriosis is a source of estrogen production. Takahashi et al. [12] measured estradiol levels in peripheral and menstrual blood of patients with endometriosis, adenomyosis, and women with normal menstrual cycles. They found no significant differences in estradiol levels in peripheral blood, whereas in the menstrual blood estradiol level was highest in women with adenomyosis, followed by the patients with endometriosis, and lowest in the patients with normal menstrual cycles. RT-PCR and Southern blot analyses have shown that P450arom is expressed in the eutopic endometria of patients with endometriosis (fig. 1) [10, 11]. P450arom protein is immunolocalized in the glandular cells in eutopic endometria of patients with endometriosis (fig. 2B), adenomyosis, and leiomyomas [11].

It has been under debate as to whether normal endometrium of women of reproductive age is capable of estrogen biosynthesis. A number of investigators [13–19] have reported the presence of aromatase activity in normal endometrium. By contrast, several groups reported the lack of aromatase activity [11, 20, 21], mRNA expression [11, 22], or protein expression [11] in normal endometrium. Careful review of the previous reports, however, shows that the endometrial specimens defined as 'normal' endometria have been obtained mostly by hysterectomy conducted for various diseases including leiomyoma, adenomyosis, and endometriosis. The specimens were not necessarily from disease-free uteri. To clarify the controversy, we strictly defined as disease-free endometrium only as eutopic endometrium obtained from normal menstruating patients with cervical cancer in situ, but no oth-

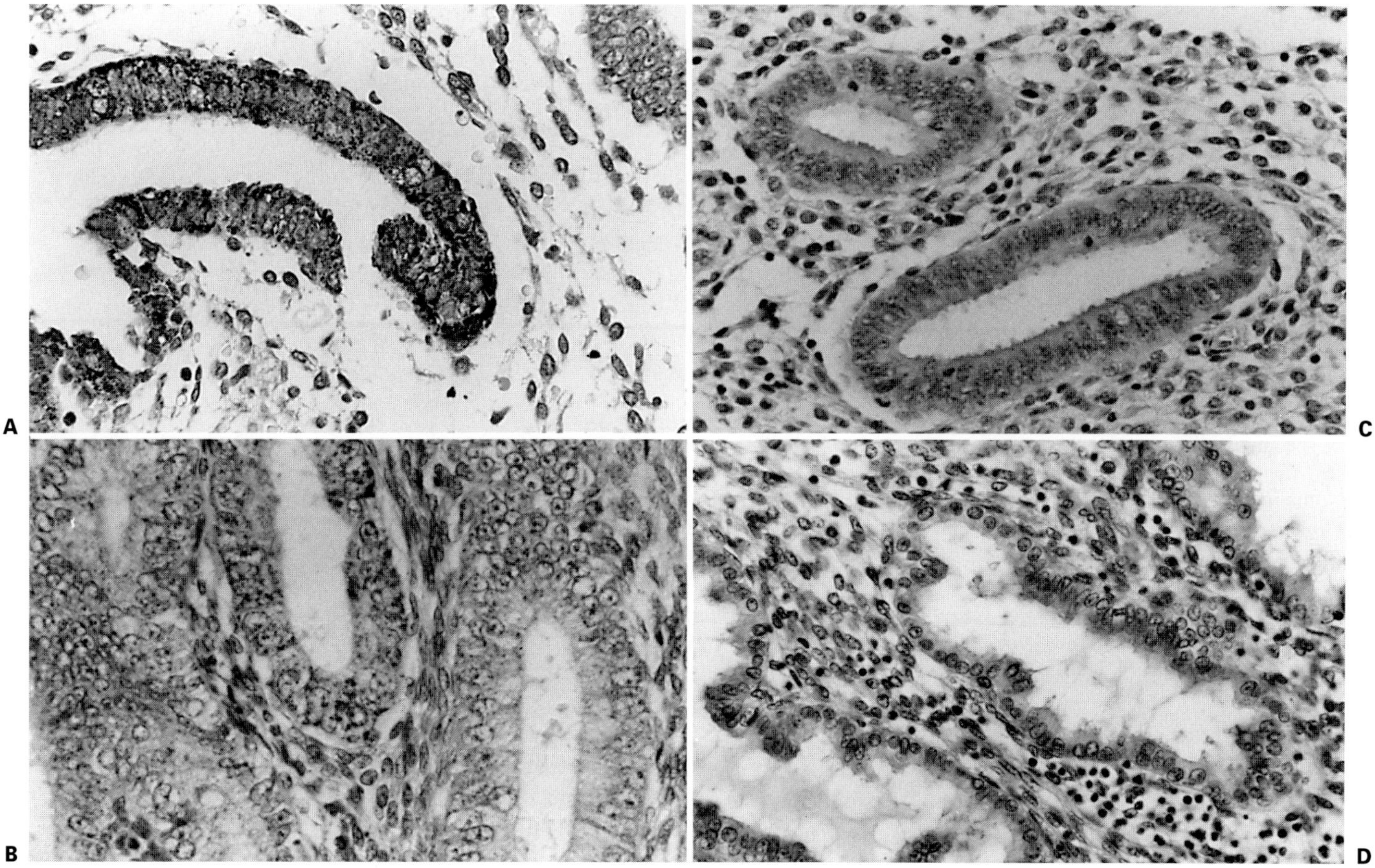

Fig. 2. Immunohistochemical localization of aromatase cytochrome P450 (P450arom) in endometriotic implant in a peritoneal vesicle implant (**A**), eutopic endometrium obtained from a patient with endometriosis (**B**), and the eutopic endometria in the proliferative phase (**C**) and secretory phase (**D**) obtained from patients with cervical carcinoma in situ but without any other gynecologic disease. Original magnification, × 100.

er gynecologic diseases, and examined P450arom expression. P450arom transcripts were not detected in disease-free endometrial specimens (fig. 1) [11]. Similarly, P450arom protein was not detected in the disease-free endometria from the proliferative phase (fig. 2C) or secretory phase (fig. 2D) [11]. The results indicate that estrogen metabolism in the eutopic endometrium appears to be remarkably different in women with endometriosis compared to disease-free women, although the histologic features of the endometria are similar in these two groups.

In the 1980s numerous studies [13–19] were done on the regulation of aromatase activity in eutopic endometrium obtained from patients with benign diseases including endometriosis, adenomyosis, and leiomyomas, although they assumed such endometrial specimens to be 'normal'. In the primary culture system of endometrium obtained from benign diseases, progestogens were potent stimulators of aromatase activity and the stimulation was inhibited by the addition of RU 486, suggesting that the stimulation was mediated via the progesterone receptors [14]. The concomitant addition of estrogens further increased the stimulation by progestogens [14]. Insulin also stimulated endometrial aromatase [17]. Recently, Noble et al. [23] showed that in the culture system of endometriotic cells derived from ovarian chocolate cysts aromatase was stimulated by prostaglandin E_2, one of the factors produced in the local level of endometriosis, suggesting that prostaglandin E_2 may stimulate local estrogen production, which thus stimulates the growth of endometriosis. In contrast, no stimulation was observed by the cytokines

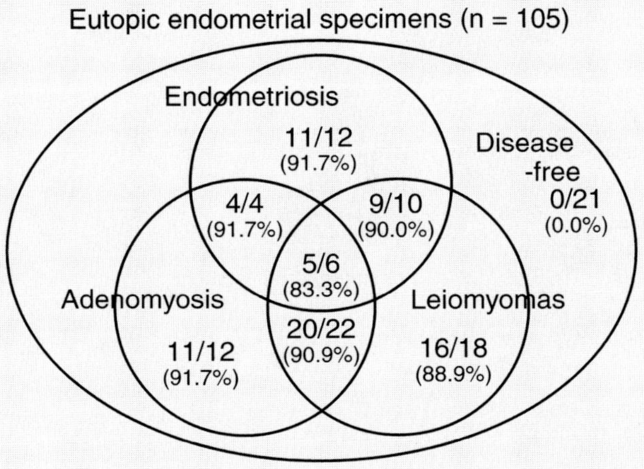

Fig. 3. Positive ratios for aromatase cytochrome P450 (P450arom) immunostaining in diseased or disease-free endometrial biopsy specimens. The overlapped areas represent the complication of multiple diseases. Numbers represent the positive immunostaining per number of cases (%) with the cutoff value for the H-score at 20.

including interleukin-1β (IL-1β), IL-2, IL-6, IL-11, IL-15 or tumor necrosis factor-α, that were increased in the peritoneal fluids of patients with endometriosis [23]. However, the mechanism by which P450arom is expressed in the eutopic endometrium of estrogen-dependent diseases is still unknown and is a question yet to be solved.

To understand the mechanism in which aromatase is present both in the ectopic glands and eutopic endometrium of patients with endometriosis, two hypotheses corresponding to two major theories of pathogenesis are proposed: the metastatic theory, which hypothesizes that aromatase expressed in eutopic endometrium by stimulation of an unknown factor activates transplantation of endometrial implants on the peritoneum; and the metaplastic theory, which hypothesizes that cytokines and growth factors mainly derived from macrophages in the peritoneal cavity stimulate the expression of aromatase in both eutopic and ectopic endometriotic glands. Further studies are necessary to elucidate the mechanism by which aromatase is expressed in these tissues. Knowledge gained from such studies would contribute to a better understanding of the growth mechanism and pathogenesis of the disease.

Application to a Diagnostic Test for Endometriosis

At present, laparotomy or laparoscopy is needed to establish a diagnosis of endometriosis [24]. CA-125 is a useful serologic marker that shows the progression of endometriosis and adenomyosis, however, its usefulness as a diagnostic test is limited because it is not sensitive enough to diagnose early-stage endometriosis [25]. The finding that P450arom is expressed in the eutopic endometria of patients with endometriosis, adenomyosis, and leiomyomas, but not in disease-free endometria, led us to evaluate the clinical utility of a test to diagnose the existence of endometriosis by detecting P450arom in endometrial biopsy specimens. We collected endometrial biopsy specimens from patients scheduled for laparotomy or laparoscopy and immunostained the specimens for P450arom. The distribution and intensity of immunostaining was analyzed by using a semiquantitative index designated H-score (0–300) [26]. Postoperative histologic diagnoses of endometriosis, adenomyosis and/or leiomyomas were made. Endometria obtained from patients with cervical cancer in situ but showing no other gynecological disease were defined as disease-free. The cases that were complicated with other gynecological diseases, including malignant neoplasms other than cervical carcinoma in situ, ovarian neoplasms, pelvic inflammation, and pregnancy, were excluded from the study.

Table 1. Sensitivity, specificity, and positive and negative predictive values for aromatase cytochrome P450 (P450arom) immunostaining in endometrial biopsy specimens

	Diseased*	Disease-free	Total
P450arom positive†	76	0	76
P450arom negative†	8	21	29
Total	84	21	105

Sensitivity = 91%; specificity = 100%; positive predictive value = 100%; negative predictive value = 72%.

* Grouped category of endometriosis, adenomyosis, and leiomyomas.

† The cutoff value for the H-score was set at 20.

Of 105 biopsy specimens, the H-scores for P450arom in the endometria obtained from patients with endometriosis, adenomyosis and/or leiomyomas varied widely. In contrast, all H-scores of the 21 specimens from disease-free uteri were below 20. The receiver-operating characteristic curve analysis showed a cutoff value of 20 with a sensitivity and specificity of 91% and 100%, respectively (table 1) [27]. The positive ratios for P450arom were similar over 80% in each group of endometriosis, adenomyosis and/or leiomyomas, regardless whether the group consisted of one disease or showed the complication of having two or three diseases (fig. 3) [27]. This technique may enable us to diagnose endometriosis more easily and without laparoscopy or laparotomy when used in combination with imaging diagnostic technologies, and may be used as an initial screening at outpatient infertility clinics.

Conclusion

In conclusion, endometriotic implants, like other estrogen-dependent tumors, contain both estrogen receptors and P450arom, suggesting that at a local level, endometriotic implants produce estrogens, which may be involved in tissue growth through interaction with the estrogen receptors. P450arom is also expressed in the eutopic endometria of patients with endometriosis, whereas P450arom is not expressed in the eutopic endometria of disease-free women. Examination of P450arom expression in endometrial biopsy specimens enables the physician to discriminate between the presence and absence of endometriosis, and may be used as an initial screening at outpatient infertility clinics.

References

1 Tamaya T, Motoyama T, Ohono Y, Ide N, Tsurusaki T, Okada H: Steroid receptor levels and histology of endometriosis and adenomyosis. Fertil Steril 1979;31:396–400.
2 Lessy BA, Metzger DA, McCarty KS: Immunohistochemical analysis of estrogen and progesterone receptors in endometriosis: Comparison with normal endometrium during the menstrual cycle and the effect of medical therapy. Fertil Steril 1989;51:409–415.
3 Tseng L, Mazella J, Funt MI, Mann WJ, Stone ML: Preliminary studies of aromatase in human neoplastic endometrium. Obstet Gynecol 1984;63:150–154.
4 Yamaki J, Yamamoto T, Okada H: Aromatization of androstenedione by normal and neoplastic endometrium of the uterus. J Steroid Biochem 1985;22:63–66.
5 Bulun SE, Economos K, Miller D, Simpson ER: CYP19 (aromatase cytochrome P450) gene expression in human malignant endometrial tumors. J Clin Endocrinol Metab 1994;79:1831–1834.
6 Yamamoto T, Takamori K, Okada H: Effect of aminoglutethimide on androstenedione aromatase activity in human uterine leiomyoma. Horm Metab Res 1985;17:548–549.
7 Bulun SE, Simpson ER, Word RA: Expression of the CYP19 gene and its product aromatase cytochrome P450 in human uterine leiomyoma tissues and cells in culture. J Clin Endocrinol Metab 1994;78:736–743.
8 Urabe M, Yamamoto T, Kitawaki J, Honjo H, Okada H: Estrogen biosynthesis in human uterine adenomyosis. Acta Endocrinol (Copenh) 1989;121:259–264.
9 Yamamoto T, Noguchi T, Tamura T, Kitawaki J, Okada H: Evidence for estrogen synthesis in adenomyotic tissues. Am J Obstet Gynecol 1993;169:734–738.
10 Noble LS, Simpson ER, Johns A, Bulun SE: Aromatase expression in endometriosis. J Clin Endocrinol Metab 1996;81:174–179.
11 Kitawaki J, Noguchi T, Amatsu T, Maeda K, Tsukamoto K, Yamamoto T, Fushiki S, Osawa Y, Honjo H: Expression of aromatase cytochrome P450 protein and messenger ribonucleic acid in human endometriotic and adenomyotic tissues but not in normal endometrium. Biol Reprod 1997;57:514–519.
12 Takahashi K, Nagata H, Kitao M: Clinical usefulness of determination of estradiol level in the menstrual blood for patients with endometriosis. Acta Obstet Gynaecol Jpn 1989;41:1849–1850.
13 Tseng L: Estrogen synthesis in human endometrial epithelial glands and stromal cells. J Steroid Biochem 1984;20:877–881.
14 Tseng L, Mazella J, Sun B: Modulation of aromatase activity in human endometrial stromal cells by steroids, tamoxifen and RU486. Endocrinology 1986;118:1312–1318.
15 Neulen J, Hartmann C, Breckwoldt M: Aromatase activity in monolayer cell cultures of human endometrium. Gynecol Endocrinol 1987;1:339–343.
16 Tseng L, Malbon CC, Lane B, Kaplan C, Mazella J, Dahler H, Tseng A: Progestin-dependent effect of forskolin on human endometrial aromatase activity. Hum Reprod 1987;2:371–377.
17 Randolph JF Jr, Kipersztok S, Ayers JW, Ansbacher R, Peegel H, Menon KM: The effect of insulin on aromatase activity in isolated human endometrial glands and stroma. Am J Obstet Gynecol 1987;157:1534–1539.
18 Yamamoto T, Shiroshita K, Kitawaki J, Okada H: The inductive effects of progestogens on aromatase activity in stromal cells of human uterine endometrium. J Endocrinol Invest 1989;12:201–204.

19 Huang JR, Bellino FL, Osawa Y, Tseng L: Immunologic identification of the aromatase enzyme system in human endometrium. J Steroid Biochem 1989;33:1043–1047.
20 Baxendale PM, Reed MJ, James VHT: Inability of human endometrium or myometrium to aromatize androstenedione. J Steroid Biochem 1981;14:305–306.
21 Prefontaine M, Shih C, Pan CC, Bhavnani BR: Applicability of the product isolation and the radiometric aromatase assays for the measurement of low levels of aromatase: Lack of aromatase activity in the human endometrium. J Endocrinol 1990;127:539–551.
22 Bulun SE, Mahendroo MS, Simpson ER: Polymerase chain reaction amplification fails to detect aromatase cytochrome P450 transcripts in normal human endometrium or decidua. J Clin Endocrinol Metab 1993;76:1458–1463.
23 Noble LS, Takayama K, Zeitoun KM, Putman JM, Johns DA, Hinshelwood MM, Agarwal VR, Zhao Y, Carr BR, Bulun SE: Prostaglandin E_2 stimulates aromatase expression in endometriosis-derived stromal cells. J Clin Endocrinol Metab 1997;82:600–606.
24 American Society for Reproductive Medicine: Revised endometriosis classification 1996. Fertil Steril 1997;67:815–821.
25 Colacurci N, Fortunato N, De Franciscis P, Fratta M, Cioffi M, Zarcone R, Cardone A: Serum and peritoneal CA-125 levels as diagnostic test for endometriosis. Eur J Obstet Gynecol Reprod Biol 1996;66:41–43.
26 McCarty KS Jr, Miller LS, Cox EB, Konrath J, McCarty KS Sr: Estrogen receptor analyses. Correlation of biochemical and immunohistochemical methods using monoclonal antireceptor antibodies. Arch Pathol Lab Med 1985;109:716–721.
27 Kitawaki J, Kusuki I, Koshiba H, Tsukamoto K, Fushiki S, Honjo H: Detection of aromatase cytochrome P450 in endometrial biopsy specimens as a diagnostic test for endometriosis. Fertil Steril 1999, in press.

Post-Presentation Discussion

Professor Terakawa: Thank you very much Dr. Kitawaki for these very interesting data. The paper is open for discussion please.

Professor Tsutsumi: It was very interesting. You have also studied patients with carcinoma in situ. What about the other demographic factors that might impact the results, like age and parity? In your speculation do you believe that it is genetically predetermined? In other words, before the establishment of the endometriosis, do you believe the aromatase activity is already present, or aromatase activity is up-regulated after the establishment of the endometriosis? P450 aromatase is, you know, related to, for example, dioxin story. And do you have anything to tell us about dioxin?

Dr. Kitawaki: Yes, as for the age or the parity of the patients with carcinoma in situ, as far as those patients are concerned there was no significant difference between groups in terms of age and parity except for those with fibroma. And as to age, all these patients that we have examined are in the reproductive age between age 24 and 48, and among those patients there was no age-related distribution difference. In other words, for the time being, we don't believe that it is influenced by age or parity. And the second question, that is the questions as to whether it is genetically predetermined or not, or whether this aromatase activity is up-regulated after the establishment of endometriosis or just prior to the establishment of endometriosis. Normally, the question is whether the endometrial cells express the aromatase or not in the disease-free patients and we have examined that and it was not expressed. So, it is something which should not normally be there, but it is expressed when the patient develops sudden disease. Alternatively, there is a small constitutive expression, but that is too low to have any biological activity maybe. So, in terms of the activity, it is not there in the normal tissue. And whether it is expressed before the establishment of the disease or not, in other words, this is the question very much related to the implant theory versus metaplasia theory.

For example, in eutopic endometrium, if P450 is expressed, then P450 aromatase activity stimulates the tissue so that the implant is more easily adhered to the peritoneum, or if the metaplasia theory is correct, then there is establishment of the disease before the induction of the aromatase. Well, there is no study to tell which is true yet. Therefore, this is something that we have to study in the future. As regards its relation to the dioxin story, the same thing. There are no data. Direct cause of aromatase induction, which is the agent? That is a very important question, but we don't have the answer to this question yet. But in the 1980s, many papers have been published which describe the regulation by aromatase of eutopic endometrium, but these studies were defective because they were examining the presence of aromatase and its activity in eutopic endometrium taken from patients with endometriosis. So, it is not from disease-free women and that is the eutopic endometrium patients with endometriosis, so progestogen is the potent stimulator of aromatase, whereas the other cytokines have no impact on it. Insulin does impact aromatase and it varies in the effect of dioxin on the expression of the aromatase, for example. Yes, very stringent inhibition of the estrogen receptor, however, with prolonged exposure there is maybe stimulation of endometrium growth. For example, that is something that we observed with tamoxifen. Therefore, I believe there is some link, which I would like to study in the future.

Dr. Kubota: My question is related to the former question. In patients with endometriosis, you said that aromatase is highly expressed in the eutopic endometrium. But the patients when they are cured of the disease or are being treated via GnRH agonist, then what happens? Because this will maybe tell us why aromatase expresses there.

Dr. Kitawaki: It is a very penetrating question. What I have observed is that during endocrine therapy, aromatase is gone in patients with endometriosis. But, when the disease relapses, then they come back. And as for danazol, it directly inhibits aromatase.

Dr. Suginami: My first question is related to the location of fibroma and the expression of aromatase. Do you know if there is any difference between the outside and inside? Are you talking about adenomyosis or fibroma?

Dr. Kitawaki: In the case of adenomyosis, I didn't carefully study these specimens from that point of view. You are right, there might be differences. If it is closer to the cavity, there might be higher expression. I will go back and check this question. As for fibroma, although I didn't show you today, more than 10 years ago it was shown that the level of expression is different between the core and the surface of the fibroma. It is higher in the surface.

Dr. Suginami: My question is the location of fibroma. Is it related to the expression level of aromatase?

Dr. Kitawaki: I didn't see any obvious difference in the case of aromatase. But sulfatase and sulfotransferase were examined, and in the case of these enzymes, the endometrium closer to the fibroma location had shown higher expression compared to the endometrium more remote from the fibroma location.

Dr. Ota: Two questions. First, you have used H-score and there were maybe 7 or 8 patients with false-positive results. That was with the protein level. What about messenger RNA level in those patients?

Dr. Kitawaki: Yes, I believe that RT-PCR is better probably, that it will give us more objective clues to the question. However, RT-PCR has very poor sensitivity and has too high false-negative results and that is the reason why I have selected immunohistochemistry.

Dr. Ota: The second question is a very simple question. Endometriosis and fibroma are estrogen-dependent diseases. However, one makes the nodule and one makes the different type of lesion. Why? What determines this difference between these two disease processes?

Dr. Kitawaki: Well, we cannot explain that from the steroid point of view. Maybe I am not answering your question, but when there is an active region, either pigmented in red or white active lesion, in case of fibroma also in the periphery of the fibroma there is high level of aromatase activity. But in chocolate cysts or in black lesions, aromatase activity is not there. The same applies to estrogen receptor as well. I am not answering your question, I know.

Dr. Fujimoto: What about different isozymes of aromatase? You know, you have used these antibodies, but antibodies have different recognition sites depending on the isoform and that might give you the different results with different diseases.

Dr. Kitawaki: Well, there is no isozyme detected yet in the protein level. You are talking about the splice variant or you know you are talking about untranslated exon 1 splice form. But the final protein product is identical, I believe.

Dr. Kurachi: When the patient has an estrogen-dependent disease process going on, then aromatase is produced, but why? How can you explain? Estrogen is believed to belong to the endocrine system, and there is always sufficient endocrine point of view, sufficient level of estrogen. So what is the pathophysiological value?

Dr. Kitawaki: This is my speculation. For example, it is known that in patients with breast cancer, both estrogen receptors and aromatase are expressed. In other words, the circulating level of estrogen is not sufficient for these cancer cells, and therefore, they produce for their need sufficient amount of estrogen in order to sustain their growth, I guess.

Dr. Suginami: In relation to that question, the patient with breast cancer, and if you take endometrium from patients with breast cancer and then, because that is what you did with these patients, eutopic endometrium of patients with breast cancer, aromatase activity, do you think, is expressed?

Dr. Kitawaki: Well, that is an interesting question. I would like to examine that.

Dr. Suginami: Because you are detecting aromatase in eutopic endometrium, so if there is an estrogen-dependent tumor process, then this can be a good screening test for any of those diseases. And therefore, if there is no endometriosis, no fibroma, then we can suspect breast cancer if it is positive.

Dr. Kitawaki: Thank you for a very interesting suggestion. I will examine that one.

Dr. Irahara: This local aromatase activity which you have examined, is the staining pattern similar to the excised lesion of endometriosis or fibroma? Because you said that aromatase activity is not correlated with the severity of the disease. So, do you think the pattern of staining is the same whether there is disease or not?

Dr. Kitawaki: That's right. That is exactly the case. Because there is no correlation with severity. So in patients with disease, both in normal tissue and in the diseased tissue, the pattern of staining is the same.

Dr. Nishida: Very, very interesting study. In estrogen-dependent disease additives such as endometriosis, or fibroma, H-score of aromatase was found to be high in the endometrium. Then, in the case of endometrial cancer or endometrium hyperplasia, is it also increased? Second question: You said it is not correlated with the disease stage or disease progression. Then when the patient is very mildly affected, extremely small endometriosis implant or fibroma, still you can detect the presence of disease?

Dr. Kitawaki: Yes, as for the first question related to endometrial cancer, some patients express it and others don't. So it is a mixed result. It depends on the cellular region, I guess, in the population of the same clone or the cells belonging to the same population are either stained positive or negative. So they are a focal distribution of positivity and negativity, mixed in the lesion. Whether the patient is pre- or post-menopausal and regardless of the stage or the grade of the disease. So, we have seen positive results, positive staining in some of the patients, but we cannot obtain any clinical usefulness as a diagnostic test. The second question: Can we detect minimal disease with this test? As Professor Tsutsumi pointed out, this is a question related to the mechanism of the disease establishment. With laparoscopic examination you find only one small implant and still you can sometimes detect P450. So, it seems that even with a very small implant, yes, we can detect this aromatase activity.

Session 2
Review

Gynecologic and Obstetric Investigation

Gynecol Obstet Invest 1999;48(suppl 1):29–35

Endometriosis and Free Radicals

Hirotaka Ota[a] Shinichi Igarashi[b] Junichi Hatazawa[a] Toshinobu Tanaka[a]

[a]Department of Obstetrics and Gynecology, Akita University School of Medicine, Akita and
[b]Department of Obstetrics and Gynecology, Koto General Hospital, Hachirogata, Japan

Key Words
Free radical · Nitric oxide · Endometrium · Endometriosis · Adenomyosis

Abstract
Recent studies have shown that synthase for nitric oxide or scavenger enzymes is distributed throughout the endometrium. We have reported that endothelial nitric oxide synthase, originally identified in vascular endothelial cells, is distributed in glandular epithelial cells in the endometrium, peaking in the midsecretory phase. In addition, it is known that superoxide dismutase is distributed throughout the endometrium, varying with the menstrual cycle. Yet it is not clear how these enzymes are committed in the reproductive processes. Endometriosis is often complicated by infertility and miscarriage. Of particular interest is that these enzymes are overexpressed in the disease throughout the menstrual cycle. These findings strongly suggest that excessive amounts of free radicals are produced in endometriosis.

What Are Free Radicals?

Free radicals are atoms or molecules with an unpaired electron. In an effort to become paired, those with an unpaired electron take an electron from stable atoms or molecules in an attempt to stabilize themselves. Because atoms or molecules that have had an electron taken away then themselves become unstable free radicals, they in turn take an electron from yet another atom or molecule in an attempt to stabilize themselves, causing a chain reaction.

We keep ourselves alive by taking oxygen in the air into our bodies and using the energy obtained from oxidation. Active oxygen is a term used for more active substances that are a product of oxygen. Active oxygen comprises both free radicals and non-free radicals. Examples of free radicals include superoxide (O_2^-), hydroxyl radical (•OH), and nitric oxide, of which the hydroxyl radical is the most active (fig. 1). Hydrogen peroxide (H_2O_2) is an example of a non-free radical. Non-free radicals have paired electrons, so they are not free radicals, but by nature they can very easily become free radicals. Because they cause very similar damage, they are thought to be closely associated.

Free radicals turn lipids into lipid peroxide, denature proteins, inactivate enzymes and oxidize and decompose the nucleobase of DNA. In this way they cause cellular and tissue damage, which results in a variety of diseases.

Endometriosis and Infertility

Endometriosis causes various symptoms such as dysmenorrhea, hypermenorrhea, and back pain. The disease is often associated with infertility, and is found in 21–48% of infertile [1, 2], compared with 1–5% [3] of fertile, women undergoing laparoscopic tubal ligation. It is a pop-

KARGER
Fax + 41 61 306 12 34
E-Mail karger@karger.ch
www.karger.com

0378–7346/99/0485–0029$17.50/0

Accessible online at:
www.karger.com/journals/goi

Hirotaka Ota, MD
Department of Obstetrics and Gynecology
Akita University School of Medicine
1-1-1 Hondo, Akita-city, Akita-ken, 010-8543 (Japan)
Tel. +81 18 884 6163, Fax +81 18 884 6447, E-Mail otah@obgyn.med.akita-u.ac.jp

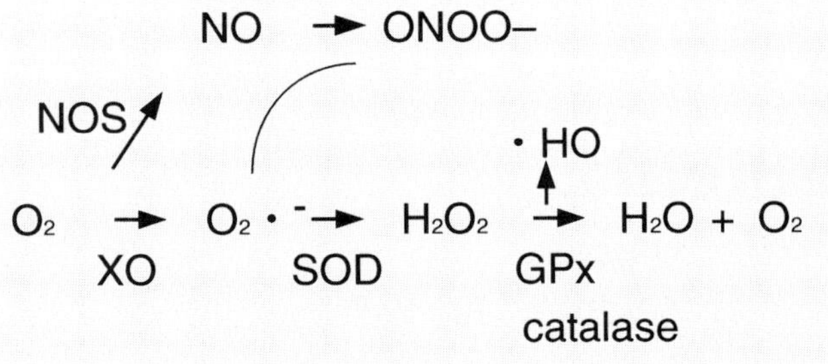

Fig. 1. Metabolic pathway of free radicals. NO = nitric oxide; NOS = nitric oxide synthase; XO = xanthine oxidase; SOD = superoxide dismutase; GPx = glutathione peroxidase.

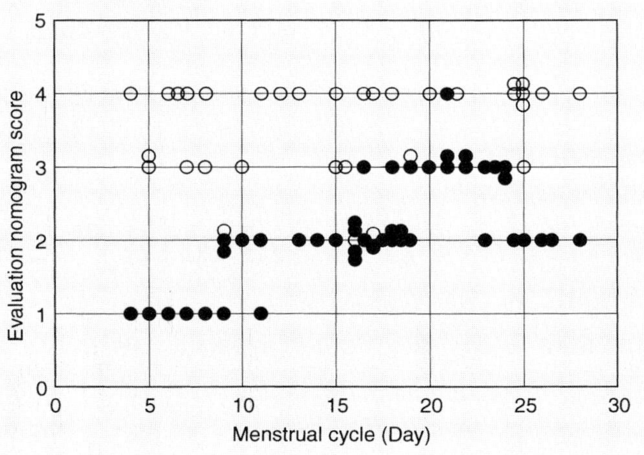

Fig. 2. Expression of endothelial nitric oxide synthase during the menstrual cycle. ● = Fertile control; ○ = endometriosis patients. Reprinted by permission from the American Society of Reproductive Medicine [Fertil Steril 1998;69:303–308].

ular belief that endometriosis is more common in infertile women than in parous women. Some reports have suggested an increased incidence of spontaneous abortion in women with endometriosis, ranging from 11–63% [4, 5], and treatment greatly reduced pregnancy wastage.

Several studies have reported an increased miscarriage rate for patients with adenomyosis [6, 7], although little attention has been paid to the possible relation between adenomyosis and infertility. Only a couple of case reports are available to date [8, 9]. Several reasons for this lack of information have been cited. The incidence of adenomyosis begins to increase from the age of the mid-thirties. It is difficult to diagnose adenomyosis before surgery because there are no pathognomonic signs, symptoms, or physical findings. However, recent developments in methods of evaluation, particularly magnetic resonance imaging, have shown the importance of adenomyosis in infertile patients [10, 11].

Thus, the question is raised: What are the mechanisms that cause infertility or induce early miscarriage in endometriosis or adenomyosis? Investigation of the endometrial environment shows that a series of immune responses are activated in the endometrium of patients with endometriosis or adenomyosis, including changes in both cellular and humoral immunity [12–21]. This abnormal immunologic environment might influence the implantation of embryos, involving free radicals as a key factor in the process.

Nitric Oxide

Nitric oxide, synthesized from L-arginine by nitric oxide synthase (NOS), has been shown to mediate various physiological processes. It has been reported that the three NOS isoenzymes, endothelial NOS, neuronal NOS, and induced NOS, are present in mammals [22]. Among the enzymes, neuronal NOS and endothelial NOS are referred to as constitutive NOS, which is constantly present in cytoplasm and produces a small amount of nitric oxide. In contrast, induced NOS is present in macrophages or endothelial cells and produces a large amount of nitric oxide.

Recent studies showed that these NOSs are present in human uterus and are involved in the reproductive processes. For example, nitric oxide regulates uterine contractility during pregnancy [23] and relaxes vascular smooth muscles [24]. Furthermore, nitric oxide controls blood pressure during pregnancy and regulates the fetomaternal circulation. There is a report that reduction of nitric oxide by a NOS inhibitor induces toxemia of pregnancy and intrauterine growth retardation [25]. Conversely, constitutive NOS and messenger RNA for the enzyme have been detected only recently in the rat and human endometrium [26–28]. Yet, the precise mechanism how NOS is involved in the endometrial function is still obscure.

We have reported that endothelial NOS is distributed throughout the human endometrium, particularly in the glandular and surface epithelium (fig. 2) [29]. The enzyme is slightly pronounced in the proliferative phase and peaks in the midsecretory phase, thus varying with the menstrual phase.

We questioned why the expression of endothelial NOS and possible liberation of nitric oxide reached a peak in the midsecretory phase. During this phase, endometrium shows marked histologic changes under the influence of progesterone. In particular, the stroma is abundant and becomes edematous. Thus, nitric oxide might be involved in stromal edematization. In addition, nitric oxide might participate in the regulation of uterine contraction [30]. In normal fertile women, contractions in the subendometrial myometrium vary with the phases of the menstrual cycle [31]. The contractions increase throughout the proliferative and periovulatory phases, and decrease in the secretory phase. From this viewpoint, nitric oxide synergistic with progesterone might relax uterine contraction in the secretory phase in a paracrine fashion.

Endometriosis and Nitric Oxide

The expression of endothelial NOS in endometriosis as well as in adenomyosis is persistently marked when compared to control levels throughout the menstrual cycle [29]. It is true that endometrial development seriously affects embryo implantation, and inconsistent development between endometrium and embryo would impede embryo implantation.

This raised the question: How does an excessive supply of nitric oxide damage fecundity in endometriosis and adenomyosis? The first possibility is that nitric oxide may disturb peristaltic uterine contraction in a paracrine fashion. Leyendecker and his colleagues [32] reported that the uterine hyperperistalsis and dysperistalsis in patients with endometriosis are significantly different from those of controls, which probably leads to disturbed sperm transport and reduced fertility. Another possibility is that either an excess or a deficiency of nitric oxide damages early embryonic development. It has been reported that injection of nitric oxide donor and NOS inhibitor into the amniotic cavity resulted in malformed embryos in rats [33]. These facts suggest the importance of optimal levels of nitric oxide in the endometrium for normal embryonic development.

Examination of the endometrial environment in endometriosis shows that various immune parameters differ widely from those in normal fertile women [13–19]. These abnormal immune responses might eventually stimulate macrophages and/or endometrial cells to persistently produce a large amount of nitric oxide and inhibit implantation. Endothelial NOS is activated by estrogens [34, 35] and cytokines such as transforming growth factor-β [36]. In contrast, the enzyme is inhibited by tumor necrosis factor-α [37]. It is notable that endometrial cells can secrete these cytokines and the host receptors for them [38]. These facts suggest that various cytokines secreted from endometrial cells, immune cells, or macrophages stimulate endothelial NOS to liberate nitric oxide.

Superoxide

It has been known that superoxide is involved in the reproductive process. For example, superoxide dismutase (SOD), an enzyme that catalyzes the dismutation of superoxide to oxygen and hydrogen peroxide, is widely distributed in the ovary [39]. The enzyme is known to inhibit ovulation [40].

It is known that SOD is located in human endometrium. Narimoto et al. [41] reported the expression of copper, zinc-SOD (Cu, Zn-SOD) in glandular and surface epithelia, and that changes in expression occur during the menstrual cycle. Similar expression was reported by Sugino et al. [42], in which enzyme activities reached a peak in the midsecretory phase. The expression of SOD in the secretory phase suggests a possible involvement of superoxide in embryonic growth or implantation. However, few reports have been available on the role of free radicals.

Oxygen tension in the fallopian tube is markedly lower than that in the atmosphere [43]. This implies that scavenger enzymes for free radicals are critical for the development of embryos. In fact, there is a report that describes the distribution of SOD in the human fallopian tube [44].

Supporting this finding are several studies that indicate that free radicals can cause embryonic arrest and cell death [45, 46]. That is, mammalian preimplantation embryos exhibit suboptimal or retarded development. In rodents, proliferation of fertilized eggs in vitro from the two-cell stage to the four-cell stage is inhibited [45, 46]. Furthermore, a large number of eggs can proceed to the four-cell stage when embryos are cultured in medium alone under a reduced oxygen atmosphere or with the addition of SOD [47]. These phenomena imply that the embryos are exposed under oxygen tension (i.e., free radicals).

Endometriosis and Superoxide

We first demonstrated the exaggerated expression of Cu, Zn-SOD, and manganese SOD (Mn-SOD) in the eutopic endometrium in endometriosis and adenomyosis (fig. 3) [48]. Ishikawa et al. [49] reported the expression of Mn-SOD in ectopic endometriotic tissues in adenomyosis. Of importance is the fact that superoxide anion reacts with nitric oxide and is metabolized into another oxidant, peroxynitrite [50]. Thus, these findings suggest a potential co-commitment of superoxide and nitric oxide in reproductive failure, such as infertility and/or miscarriage in endometriosis and adenomyosis.

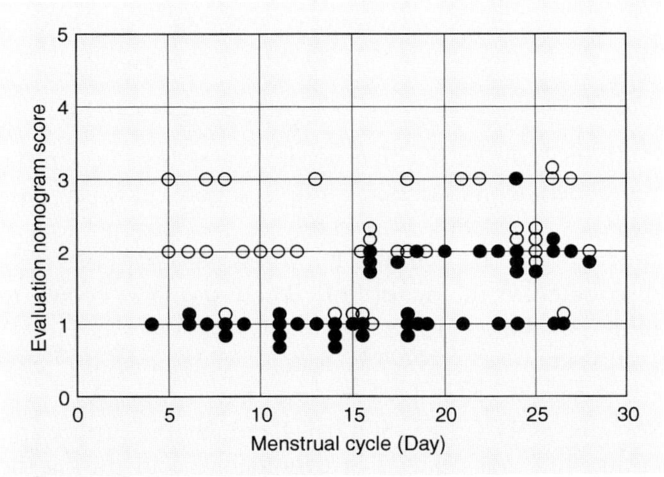

Fig. 3. Expression of manganese superoxide dismutase during the menstrual cycle. ● = Fertile control; ○ = endometriosis patients. Reprinted by permission from the American Society of Reproductive Medicine [Fertil Steril 1999;72:129–134].

This raises the following question: What is the mechanism that stimulates the expression of SOD in endometriosis and adenomyosis? It is well known that there are various abnormal immune phenomena in endometriosis and adenomyosis. Immunohistochemical studies have shown a series of cellular abnormalities in the endometrial stroma in these diseases. A number of reports indicate that levels of cytokines such as interleukin-1 or tumor necrosis factor are increased in the peritoneal fluid in endometriosis [51]. It is important to note that these cytokines are inducers of Mn-SOD [52, 53]. These facts suggest that it is quite likely that various cytokines secreted from endometrial cells, immune cells, or macrophages stimulate the expression of SOD.

Free Radicals and Infertility

The final question is: Do free radicals participate in fecundity? A recent study indicates that the local administration of NOS inhibitor during the preimplantation period markedly reduced the pregnancy rate [54]. We have also shown that in pregnant mice the administration of nitric oxide donor or NOS inhibitor during the similar period reduced the pregnancy rate in a dose-dependent manner [55]. In these mice, histologic analysis of uteri on

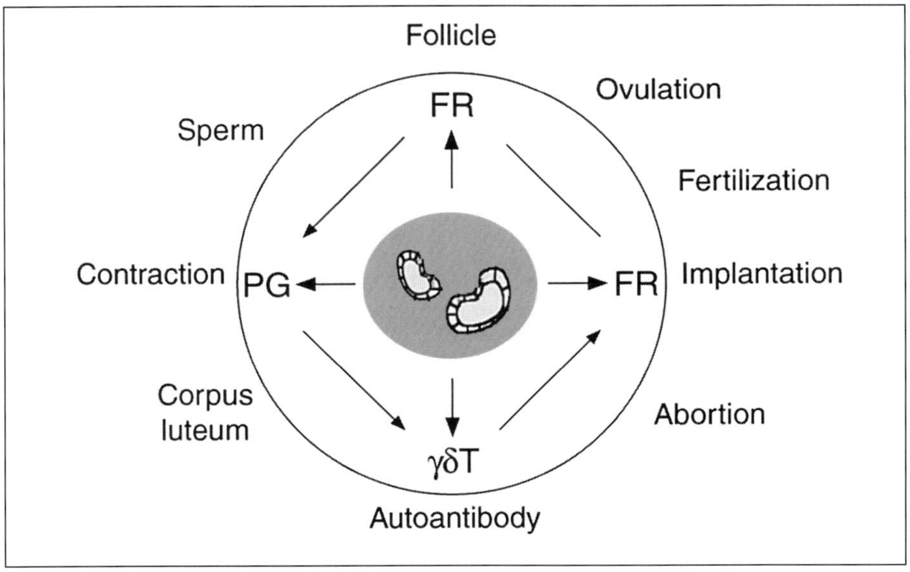

Fig. 4. Hypothesis: possible involvement of free radicals in reproduction. FR = Free radical; PG = prostaglandin; γδT = γδT cell.

day 5 of gestation after the treatment suggested impaired intercellular reactions and retarded 'decidualization' of stromal cells or defective function of decidualized cells. There is another report that indicates optimal levels of nitric oxide is critical for embryo implantation in rats [56].

From these findings, it is concluded that free radicals appear to participate intimately in implantation, and the optimal levels of free radicals are of a key importance during the process (fig. 4).

Acknowledgements

This work was partially supported by the Scientific Grant No. 10671512 from the Ministry of Education, Japan.

References

1 Hasson HM: Incidence of endometriosis in diagnostic laparoscopy. J Reprod Med 1976;16:135–138.
2 Drake TS, Grunert GM: The unsuspected pelvic factor in the infertility investigation. Fertil Steril 1980;34:27–31.
3 Berger GS: Epidemiology of endometriosis; in Nezhat CR, Berger GS, Nezhat FR, Buttram VC Jr, Nezhat CH (eds): Endometriosis. New York, Springer, 1995, pp 1–7.
4 Metzger DA, Olive DL, Stohs GF, Franklin RR: Association of endometriosis and spontaneous abortion: Effect of control group selection. Fertil Steril 1986;45:18–22.
5 Daya S: Endometriosis and spontaneous abortion; in Coutinho EM, Spinola P, de Moura LH (eds): Progress in the management of endometriosis. New York, Parthenon, 1995, pp 61–68.
6 Parazzini F, Vercellin P, Panazza S, Chatenoud L, Oldani S, Crosignani G: Risk factors for adenomyosis. Hum Reprod 1997;12:1275–1279.
7 Kano T, Furudono M, Nabetani H, Ishii M, Kanda T, Ueki M: The incidence of endometriosis and adenomyosis in patients with habitual abortion in relation to immunological abnormalities (in Japanese). Jpn J Fertil Steril 1997;42:113–118.
8 Hirata JD, Moghissi KS, Ginsburg KA: Pregnancy after medical therapy of adenomyosis with a gonadotropin-releasing hormone agonist. Fertil Steril 1993;59:444–445.
9 Honore LH, Cumming DC, Dunlop DL, Scott JZ: Uterine adenomyoma associated with infertility. A report of three cases. J Reprod Med 1988;33:331–335.
10 McCarthy S: Magnetic resonance imaging in the evaluation of infertile women. Magn Reson Q 1990;6:239–249.
11 Woodward PJ, Wagner BJ, Farley TE: MR imaging in the evaluation of female infertility. Radiographics 1993;13:293–310.
12 Ota H, Maki M, Shidara Y, Kodama H, Takahashi H, Hayakawa M, Fujimori R, Kushima T: Effects of danazol at the immunological level in patients with adenomyosis, with special reference to autoantibodies: a multi-center cooperative study. Am J Obstet Gynecol 1992;167:481–486.
13 Ota H, Igarashi S: HLA-DR expression in endometriotic tissue in patients with endometriosis and adenomyosis. Fertil Steril 1993;60:834–838.
14 Ota H, Igarashi S, Hayakawa M, Matsui T, Tanaka H, Tanaka T: Effect of danazol on the immunocompetent cells in the eutopic endometrium in patients with endometriosis: a multicenter cooperative study. Fertil Steril 1996;65:545–551.
15 Ota H, Igarashi S, Tanaka T: Expression of γδT cells and adhesion molecules in endometriotic tissue in patients with endometriosis and adenomyosis. Am J Reprod Immunol 1996;35:477–482.
16 Ota H, Tanaka T: Integrin adhesion molecules in the endometrial glandular epithelium in patients with endometriosis or adenomyosis. J Obstet Gynaecol Res 1997;23:485–491.
17 Ota H, Igarashi S, Hatazawa J, Toshinobu T: Distribution of heat shock proteins in eutopic and ectopic endometrium in endometriosis and adenomyosis. Fertil Steril 1997;68:23–28.
18 Ota H, Igarashi S, Hatazawa J, Tanaka T: Is adenomyosis an immune disease? Hum Reprod Update 1998;4:360–367.
19 Ota H, Igarashi S, Hatazawa J, Tanaka T: Immune reaction. Understanding endometriosis. New York, Parthenon, 1999, pp 143–148.
20 Lessey BA, Castelbaum AJ, Sawin SW, Buck CA, Schinnar R, Bilker W, Strom BL: Aberrant integrin expression in the endometrium of women with endometriosis. J Clin Endocrinol Metab 1994;79:643–649.
21 Lessey BA, Damjanovich L, Coutifaris C, Castelbaum A, Albelda SM, Buck CA: Integrin adhesion molecules in the human endometrium. Fertil Steril 1992;62:497–506.
22 Knowles RG, Moncada S: Nitric oxide synthases in mammals. Biochem J 1994;298:249–258.
23 Yallampalli C, Byam-Smith M, Sharon MS, Garfield RE: Steroid hormones modulate the production of nitric oxide and cGMP in the rat uterus. Endocrinology 1994;134:1971–1974.
24 Ignarro LB, Harbison RG, Wood KS, Kadowitz PJ: Activation of purified soluble guanylate cyclase by endothelium-derived relaxing factor from intrapulmonary artery and vein in stimulation by acetylcholine, bradykinin and arachidonic acid. J Pharmacol Exp Ther 1986;237:893–900.
25 Yallampalli C, Garfield RE: Inhibition of nitric oxide synthesis in rats during pregnancy produces signs similar to those of preeclampsia. Am J Obstet Gynecol 1993;169:1316–1320.
26 Telfer JF, Lyall F, Norman JE, Cameron IT: Identification of nitric oxide synthase in human uterus. Hum Reprod 1995;10:19–23.
27 Tseng L, Zhang J, Peresleni TYu, Goligorsky MS: Cyclic expression of endothelial nitric oxide synthase mRNA in the epithelial glands of human endometrium. J Soc Gynecol Invest 1996;3:33–38.
28 Telfer JF, Irvine GA, Kohnen G, Campbell S, Cameron IT: Expression of endothelial and inducible nitric oxide synthase in non-pregnant and decidualised human endometrium. Mol Hum Reprod 1997;3:69–75.
29 Ota H, Igarashi S, Hatazawa M, Tanaka T: Endothelial nitric oxide synthase in the endometrium during the menstrual cycle in patients with endometriosis and adenomyosis. Fertil Steril 1998;69:303–308.
30 Yallampalli C, Garfield RE, Byam-Smith M: Nitric oxide inhibits uterine contractility during pregnancy but not during delivery. Endocrinology 1993;133:1899–1902.
31 Lyons EA, Taylor PJ, Zheng XH, Ballard G, Levi CS, Kredentser JV: Characterization of subendometrial myometrial contractions throughout the menstrual cycle in normal fertile women. Fertil Steril 1991;55:771–774.
32 Leyendecker G, Kunz G, Wildt L, Beil D, Deininger H: Uterine hyperperistalsis and dysperistalsis as dysfunctions of the mechanism of rapid sperm transport in patients with endometriosis and infertility. Hum Reprod 1996;7:1542–1551.
33 Lee QP, Juchau MR: Dysmorphogenic effects of nitric oxide (NO) and NO synthase inhibition: Studies with intra-amniotic injections of sodium nitroprusside and NG- monomethyl-*L*-arginine. Teratology 1994;49:452–464.
34 Rao VSN, Chaves MC, Ribeiro RA: Nitric oxide synthase inhibition and the uterotrophic response to oestrogen in immature rats. J Reprod Fertil 1995;105:303–306.
35 Chaves MC, Ribeiro RA, Rao VSN: Possible involvement of nitric oxide in estrogen-induced uterine edema in the immature rat. Braz J Med Biol Res 1993;26:853–857.

36 Inoue N, Venema RC, Sayegh HS, Ohara Y, Murphy TJ, Harrison DG: Molecular regulation of the bovine endothelial cell oxide synthase by transforming growth factor-beta 1. Arterioscler Thromb Vasc Biol 1995;15:1255–1261.

37 Mohamed F, Monge JC, Gordon A, Cernacek P, Blais D, Stewart DJ: Lack of role for nitric oxide (NO) in the selective destabilization of endothelial NO synthase mRNA by tumor necrosis factor-alpha. Arterioscler Thromb Vasc Biol 1995;15:52–57.

38 Tabibzadeh S: Cytokines and the hypothalamic-pituitary-ovarian endometrial axis. Hum Reprod 1994;9:947–967.

39 Laloraya M, Pradeep KG, Laloraya MM: Changes in the levels of superoxide anion radical and superoxide dismutase during the estrous cycle of *Rattus norvegicus* and induction of superoxide dismutase in rat ovary by lutropin. Biochem Biophys Res Commun 1988;57:146–153.

40 Sato EF, Kobuchi H, Edashige K, Takahashi M, Yoshioka T, Utsumi K, Inoue M: Dynamic aspects of ovarian superoxide dismutase isozymes during the ovulatory process in the rat. FEBS Lett 1992;303:121–125.

41 Narimoto K, Noda Y, Shiotani M, Tokura T, Goto Y, Takakura K, Mori T, Fujimoto K, Ogawa K: Immunohistochemical assessment of superoxide dismutase expression in the human endometrium throughout the menstrual cycle. Acta Histochem Cytochem 1990;23:487–497.

42 Sugino N, Shimamura K, Takiguchi S, Tamura H, Ono M, Nakata M, Nakamura Y, Ogino K, Uda T, Kato H: Changes in activity of superoxide dismutase in the human endometrium throughout the menstrual cycle and in early pregnancy. Hum Reprod 1996;11:1073–1078.

43 Mastroianni L Jr, Jones R: Oxygen tension within the rabbit fallopian tube. J Reprod Fertil 1965;9:99–102.

44 Narimoto K, Noda Y, Shiotani M, Natsuyama S, Mori T, Fujimoto K, Ogawa K, Kim YC: Role of superoxide dismutase in the fallopian tube function: Immunohistochemical assessment of superoxide dismutase in the fallopian tube. Acta Histochem Cytochem 1991;24:85–91.

45 Yanagimachi R, Chang MC: In vitro fertilization of golden hamster ova. J Exp Zool 1964;156:361–376.

46 Whiten WK: Culture of tubal ova. Nature 1957;179:1081–1082.

47 Noda Y, Matsumoto H, Umaoka Y, Tatsumi K, Kishi J, Mori T: Involvement of superoxide radicals in the mouse two-cell block. Mol Reprod Dev 1991;28:356–360.

48 Ota H, Igarashi S, Hatazawa J, Tanaka T: Immunohistochemical assessment of superoxide dismutase expression in the endometrium in endometriosis and adenomyosis. Fertil Steril 1999;69:303–308.

49 Ishikawa M, Nakata T, Yaginuma Y, Nishiwaki K, Goishi K, Saitoh S: Expression of superoxide dismutase (SOD) in adenomyosis. Am J Obstet Gynecol 1993;169:730–734.

50 Padmaja S, Huie RE: The reaction of nitric oxide with organic peroxyl radicals. Biochem Biophys Res Commun 1993;195:539–544.

51 Arici A, Oral E: The peritoneal environment in endometriosis; in Diamond MP, Osteen KG (eds): Endometrium and Endometriosis. Malden, Blackwell Science, 1997, pp 161–173.

52 Wong GHW, Goeddel DV: Induction of manganous superoxide dismutase by tumour necrosis factor: Possible protective mechanisms. Science 1988;242:941–944.

53 Ono M, Kohda H, Kawaguchi T, Ohhira M, Sekiya C, Namiki M, Takeyasu A, Taniguchi N: Induction of Mn-superoxide dismutase by tumor necrosis factor, interleukin-1 and interleukin-6 in human hepatoma cells. Biochem Biophys Res Commun 1992;182:1100–1107.

54 Biswas S, Kabir SN, Pal AK: The role of nitric oxide in the process of implantation in rats. J Reprod Fertil 1998;114:157–161.

55 Ota H, Igarashi S, Hatazawa M, Tanaka T: Significance of optimal levels of nitric oxide in fecundity in mice (in Japanese). Endometriosis Kenkyukaishi 1998;19:107–110.

56 Chwalisz K, Winterhager E, Thienel T, Garfield RE: Synergistic role of nitric oxide and progesterone during the establishment of pregnancy in the rat. Hum Reprod 1999;14:542–552.

Post-Presentation Discussion

Professor Hoshiai: Thank you very much, Dr. Ota, for your interesting lecture. Does anyone have questions or comments?

Dr. Kubota: I thank you very much for your very interesting lecture. I have always admired your research activities. So, in the case of endometriosis, why does eNOS expression rise? There is a balance of cytokines and growth factors stressed. There are many factors. In that case lipopolysaccharide (LPS) is also involved. eNOS, I think, has evolved here. Have you studied iNOS? What was the relationship with eNOS?

Professor Ota: Yes, I was initially interested in iNOS. At the protein level we studied the expression, but as you know, iNOS is very difficult to stain at the protein level. In fact eNOS was rather easy to see the expression. The protein level and messenger RNA level in both ways it was easy to see the clean staining with eNOS. I understand that there is a possible involvement of iNOS. However, iNOS expression might be discussed differently. For instance, when there is shock, ischemic condition or severe infectious situations we also see the induction of iNOS expression. You also have different data and we don't have a clear conclusion. I really can't answer your question.

Dr. Kubota: According to our opinion, at the molecular level we manage to check the iNOS expression. We compared it with eNOS expression at the molecular level. At the Northern blot level we managed to see the expression. With the Northern blot histologically we did not see cyclic difference. But, with iNOS using, as there was with RT-PCR, we managed to see very subtle expression. By adding cytokines such as interleukin-β and interferon-γ, the stromal cells were stained using Southern blot. We have those data. Therefore, just like eNOS, iNOS seems to have deep involvement with the cytokine activities during the early gestation period it becomes expressed and as you said, it probably has a negative impact on the implantation process.

Professor Ota: Yes, I understand that is highly possible. But, as I mentioned, eNOS and iNOS, there is a physiologically different role. eNOS is present in the vascular endothelium and the physiological role is important in a normal condition. On the other hand, iNOS, as I said, is not found in a normal condition. It appears only when something abnormal occurs in the body. When we think about the human reproductive process, it is a very natural, normal process, whether it may be ovary or the endometrium in a one-month cycle. It is a normal physiological cycle change. Considering that background, eNOS and iNOS, what is the emphasis and weight of their importance? We have to understand the balance. I agree with you.

Dr. Kubota: I have a problem about the secretory phase and the cytokines like macrophage in the early gestation period. The immunoreactive cytokine expression is upregulated. So the impact of cytokines becomes more important then, rather than eNOS, iNOS seems to be playing a more critical role. In that case, the relation with immunoreactive cells is an interesting area.

Professor Taketani: I think at the time being the correlation and the desired relationship between the elevation in the eNOS expression and negative and the endometriosis it is still to be examined, okay? You are relieved with my comment. And, my question is, I think your presentation suggests that the eNOS expression in endometriosis may be responsible for the low compromise of the reproductive function. If this is the case, we can suggest to the speaker that the amount of eNOS expression in endometriosis is different between the women where reproductive function is proven and infertile women. Do you have any data to suggest the eNOS difference?

Professor Ota: It is a very interesting question, but the cases I have presented here include both fertile and infertile women, so I don't have much data about the outcome of fertile women and endometriosis because I can't tell you exactly which one is higher and lower.

Professor Taketani: Then, no comment about the correlation of the relationship between the eNOS and endometriosis?

Professor Ota: Dr. Kitawaki mentioned that endometriosis is estrogen-dependent. So I think the primary inducer would be estrogen.

Professor Osteen: I have a similar comment just trying to reconcile in my mind a little bit the data with higher aromatase activity and so theoretically have higher local estrogen concentrations and you are correlating free radicals higher in endometriosis and it is associated with implantation sites. I am just wondering why we have such a reduced fertility with endometriosis and correlate with these results? Because you would almost think that these would be events that would not necessarily disrupt pregnancy to the extent that is seen with this disease. Do you have any sense of why in women with endometriosis, in terms of the free-radical expression you have seen, you would have such a reduced fertility just based on the free-radical regulation?

Professor Ota: Do you need to know why free radicals cause infertility? That's why I did mouse experiments. I found that free radicals are produced in the endometrium. So the excessive amount of free radicals is a problem, I think. For example, Dr. Lessey in the United States mentioned that some kind of integrin expression is very low. So, I think, for example, nitric oxide may inhibit the expression of adhesion molecules. That is one possibility.

Dr. Suginami: Thank you very much for your beautiful presentation. You showed those iNOS and SOD levels are higher in the endometrium in endometriosis than in the normal endometrium. I would like to ask you about the removal of endometriosis. If you remove endometriosis, for instance by laparoscopy, can you reduce the level of NOS and SOD in the endometrium?

Professor Ota: Well, it is a very interesting question. I think the expression may not change, because as I mentioned a little bit in the last slide, the immune system in endometriosis is already changed. It means that T cells already have memory of these abnormal antigens. So, the expression may not change.

Dr. Momoeda: According to your experiment, you may speculate that the higher level of free radicals is responsible for the implantation inhibition. But there are many papers which reported that pregnancy rate, if we perform the IV FET, offers no difference between a woman with endometriosis and a woman without endometriosis. So, how can you explain this part?

Professor Ota: It is a very important question. I am talking about the natural cycle, which means that we are not using the drugs for ovulation induction. And when we administer hormonal medicine like hMG, estrogen levels rise to about 5 to 10 times higher than normal, which means the environment of these enzymes becomes totally different. So it is a different problem, between the fertile, unstimulatory cycle and stimulatory cycle. And I agree with you that many groups state that there is no difference in the pregnancy rate between the non-endometriosis and in these groups. But also there are several reports saying that the pregnancy rate in endometriosis is lower than in the patients without endometriosis.

Session 2
Review

Gynecologic and
Obstetric Investigation

Gynecol Obstet Invest 1999;48(suppl 1):36–44

Immunohistochemical Study of Angiogenic Factors in Endometrium and Endometriosis

Akira Fujishita Atsuko Hasuo Khaleque Newaz Khan Hideaki Masuzaki
Hisayoshi Nakashima Tadayuki Ishimaru

Department of Obstetrics and Gynecology, Nagasaki University School of Medicine, Nagasaki, Japan

Key Words
Endometriosis · Endometrium · Angiogenesis · PCNA · VEGF · Endoglin

Abstract
The proliferative activity and angiogenic factors of pelvic endometriosis and uterine endometrium and their correlation with different pigmented lesions are important throughout the menstrual cycle. The proliferative activity and/or expression of angiogenic factors appears to be different between endometrium and endometriosis, and also different in each pigmented lesion. Immunohistochemical studies using computerized image analysis have shown that in normal endometrium, the PCNA index shows cyclic variation, but no cyclic change is observed in endometriosis. In the pelvic peritoneum, the PCNA index is higher in red lesions compared to black and white lesions. There is no difference in VEGF expression among different pigmented lesions. However, a cyclic variation of VEGF concentration was found in the peritoneal fluid of patients with endometriosis. As a newly determined angiogenic factor, the number of endoglin-positive cells and the mean endothelial area were higher in red lesions than in black or white lesions. These results suggest that the pathogenesis and activity of endometriosis possibly depends on internal angiogenesis. The cellular activity of endometriotic lesions may not necessarily coincide with the regulation of angiogenesis, and may not be synchronized in each pigmented pelvic lesion of endometriosis.

Introduction

There are several hypotheses concerning the pathogenesis of endometriosis such as implantation theory, lymphatic/vascular metastasis, uterotubal migration, and coelomic metaplasia theory [1]. Although no single theory can explain all cases, it is likely that some composite mechanisms are involved in the pathogenesis of endometriosis. Endometriosis usually occurs in women of reproductive age; it is seldom seen in premenarche or postmenopausal women. Endometriosis depends on ovarian steroid hormones; estrogen appears to be the most important hormone for the progression and maintenance of endometriosis. Recently, new criteria for pigmented lesions in pelvic endometriosis have been established. Lesions are macroscopically classified as red, black, or white lesions, but the biological significance of these lesions has not been determined [2]. Angiogenesis is also important

KARGER
Fax + 41 61 306 12 34
E-Mail karger@karger.ch
www.karger.com

0378–7346/99/0485–0036$17.50/0

Accessible online at:
www.karger.com/journals/goi

Dr. Akira Fujishita
Department of Obstetrics and Gynecology
Nagasaki University School of Medicine
1-7-1 Sakamoto, Nagasaki 852-8501 (Japan)
Tel. +81 958 49 7363, Fax +81 958 49 7365, E-Mail fujisita@net.nagasaki-u.ac.jp

to cell attachment, maintenance, progression, and cell growth. However, a few reports revealed a difference in these pigmented lesions according to proliferation or expression of angiogenic factors [3–6].

Experimental Design for the Role of Angiogenesis in Endometriosis

We performed a study to investigate the proliferative activity of eutopic endometrium, endometriosis and their possible relation with the progression of angiogenesis in different endometriotic lesions. The background of our study is based on the angiogenesis-related overexpression of estrogen and progesterone receptors in endometriosis, as described in our previous study [7]. For this purpose, we performed an immunohistochemical study to detect proliferating cell nuclear antigen (PCNA), tissue expression of some angiogenic factors such as vascular endothelial growth factor (VEGF) and endoglin and to examine their correlation with different pigmented lesions of endometriosis. VEGF was also measured in the peritoneal fluid of patients with endometriosis and without endometriosis. Human endoglin (CD105) is a cell-surface component of transforming growth factor-beta (TGF-β) type I receptor complex and has been found on endothelial cells of capillaries, arterioles, small arteries, and venules of endometrium and pelvic endometriosis [8].

Sample Collection

All patients who had had no hormonal treatment for 3 months before biopsy were subjected to laparoscopy or laparotomy, and pathologic diagnosis of the biopsy specimens was made retrospectively. In all patients, accurate menstrual dating could be corroborated with appropriate histologic dating according to the criteria of Gompel and Silverberg [9]. Endometrial tissues were obtained from normal parts of the uteri of 25 patients with uterine myoma or cervical cancer. The histologic dating was classified as follows: menstrual phase (n = 4), early proliferative phase (n = 5), midproliferative phase (n = 4), late proliferative phase (n = 4), early secretory phase (n = 4) and mid to late secretory phase (n = 4).

Sixty-seven endometriosis lesions were biopsied from various pigmented lesions from 34 patients with endometriosis. Among the endometriotic implants, 30 samples which histopathologically revealed both endometrial epithelium and stroma were diagnosed as typical endometriosis, and other 37 samples were excluded from this analysis and were designated as nontypical endometriosis, inclusion cyst, reactive mesothelium, connective tissue, or normal peritoneum. The cyclic distribution of the study samples was as follows: menstrual phase, n = 3; proliferative phase, n = 11; and secretory phase, n = 16, according to the history of the menstrual cycle. The macroscopic findings of endometriotic implants in peritoneal cavity were defined according to r-AFS score [10], and classified as red (n = 8), black (n = 15) and white lesions (n = 7).

Immunohistochemical staining of PCNA, VEGF and Endoglin

For the immunohistochemical study, 25 specimens of endometrium and 30 of endometriosis were prepared. Nonimmune mouse immunoglobulin (Ig) G1 antibody was used as a negative control. The sections were reacted with the first antibody (PCNA and VEGF), then the second (biotin-labeled rabbit anti-mouse IgG+ IgA+), and then washed with PBS and reacted with HRP-conjugated streptavidin for 10 min. Visualization of HRP sites was performed by using 3,3′-diaminobenzidine 4 HCl, H_2O_2, Co^{2+}, and Ni^{2+}, according to the method of Adams [11].

The CSA system was used in the immunohistochemical staining of endoglin. Visualization of binding sites was possible by using 3-amino-9-ethylcarcarbazole in *N,N*-dimethylfolmamide and 0.3% hydrogen peroxidase. After rinsing in distilled water, a light counterstaining was done in Mayer's hematoxylin.

Measurement of VEGF Concentration

A separate study population comprising women with an age range of 19–42 years, who were not taking medication and who were undergoing diagnostic laparoscopy for either infertility or clinically suspected endometriosis, were enrolled. Patients were given a diagnosis of no endometriosis or endometriosis after laparoscopic investigation and were staged based on r-AFS classification. Controls were women in whom no visible evidence of pelvic endometriosis was found. VEGF concentration in peritoneal fluid was measured by a sandwich enzyme-linked immunosorbent assay (ELISA) system.

Computerized Analysis System

Our computerized image analysis system was comprised of a light microscope (Nikon, Japan), an image grabber 24 (Neotech, Japan) with a color charge CCD camera (Sony, Japan) connected to a Mac SCOPE computer system (Mitani Corp., Japan). PCNA index was defined as number of strongly positive cells per total counted cells × 100 (%). Staining for endoglin was done

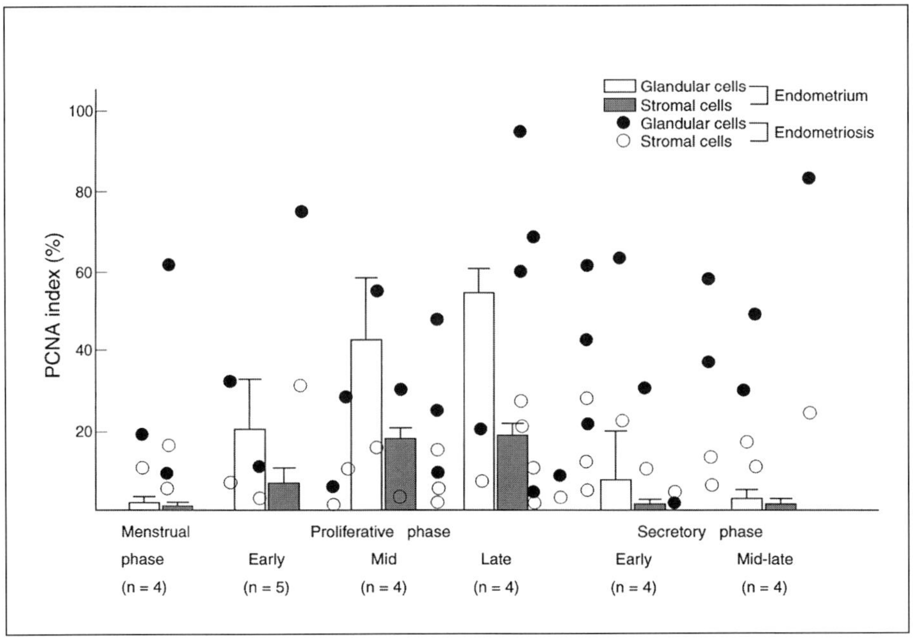

Fig. 1. Cyclic variation of the proliferating cell nuclear antigen (PCNA) index in eutopic endometrium and endometriosis. Open bar represents endometrial glandular epithelium and shaded bar shows endometrial stromal tissues in eutopic endometrium. The height of the bar represents the mean value, and the error bar indicates the SD. Closed circles denote glandular epithelium and open circles denote stromal tissues of endometriosis.

in the endothelial cells of capillaries, arterioles, and venules. A binary image was also used in this analysis. The number of microvessels was counted manually in each investigated area.

Results

PCNA Index in Endometrium and Endometriosis

In the endometrium, the glandular epithelium revealed a significantly higher PCNA index during late proliferative phase, mid proliferative phase and early proliferative phase when we compared with early secretory phase, mid-to-late secretory phase, and menstrual phase. In contrast, PCNA index of stroma in these cases was found to be lower than glandular epithelium, although a similar tendency of increasing index was observed throughout the menstrual cycle (fig. 1).

In pelvic endometriosis, no statistically significant cyclic variations of PCNA index in both glandular epithelium and stroma were observed throughout the menstrual cycle (fig. 1). When we analyzed the PCNA index among different pigmented lesions of endometriosis, we found that PCNA index of glandular epithelium appeared to be significantly higher in red lesions compared to black or white lesions. A significant difference of PCNA index was also found between black and white lesions (fig. 2).

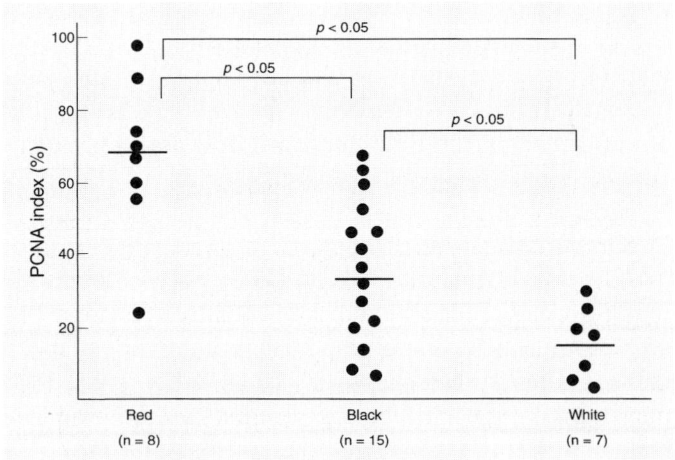

Fig. 2. Proliferating cell nuclear antigen (PCNA) index of glandular epithelium in different pigmented lesions of endometriosis. Horizontal lines represent the means.

VEGF Concentration in Peritoneal Fluid

The VEGF concentration in the peritoneal fluid of 50 patients with endometriosis was significantly higher than that of 13 controls without endometriosis. No significant difference in VEGF concentration was observed among different r-AFS stages of endometriotic implants. In patients with endometriosis, a marked cyclic variation of VEGF content in peritoneal fluid was observed. The peri-

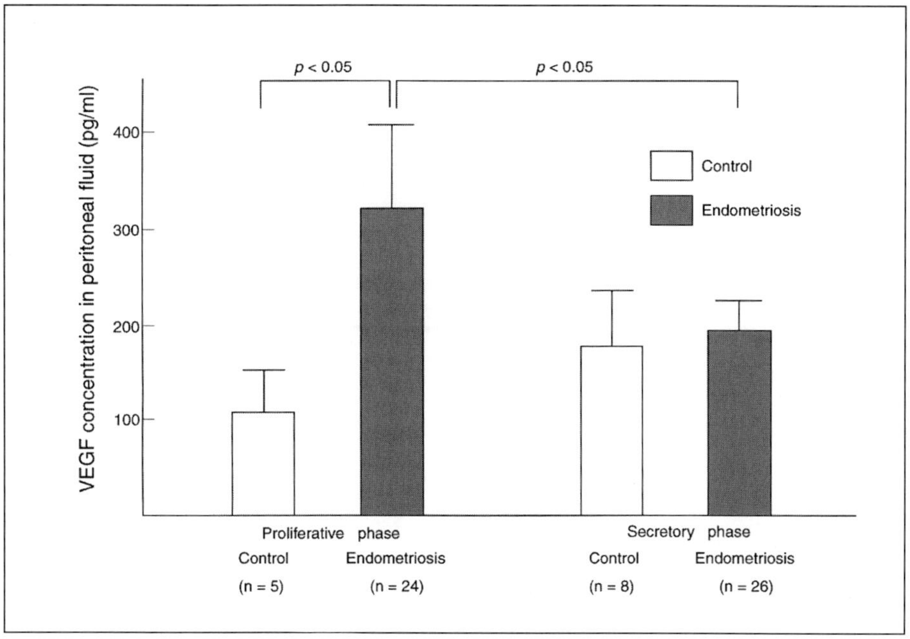

Fig. 3. Cyclic distribution of VEGF concentration in the peritoneal fluid of patients with endometriosis. Shaded bar represents patients with endometriosis and open bar represents controls. The height of the bar denotes mean value and the error bar indicates SD.

toneal fluid in the proliferative phase displayed a significantly greater concentration of VEGF than that in the secretory phase. This cyclic variation was not evident in fluid from controls (fig. 3).

Expression of VEGF in Endometrium and Endometriosis

VEGF positive cells were seen in the cytoplasm of glandular epithelium, macrophage in stromal tissues, and also in endothelial cells of stromal microvessels. In the endometrial epithelium and stroma, there were no significant cyclic changes in VEGF expression throughout the menstrual cycle. The expression of VEGF was also uniform in both epithelium and stroma in pelvic endometriosis.

Expression of Endoglin in Endometrium and Endometriosis

Most of the endoglin-positive cells were observed in the stromal tissues of endometrium and connective tissues of endometriotic pelvic peritoneum; no staining was seen in the glandular epithelium of either endometrium or endometriosis. In contrast to VEGF, the expression of endoglin was found strongly positive in red lesions and moderately to weakly positive in black lesions with a weak to sparse distribution in white lesions. According to computed analysis, there was no relationship of endoglin expression in pelvic endometriosis throughout the menstrual cycle. However, there was a significant difference in mean endothelial area occupied by endoglin in stromal cells between red and white lesions and between black and white lesions, but no remarkable difference was found between red and black lesions (fig. 4). In addition, the number of microvessels indicating neovascularization

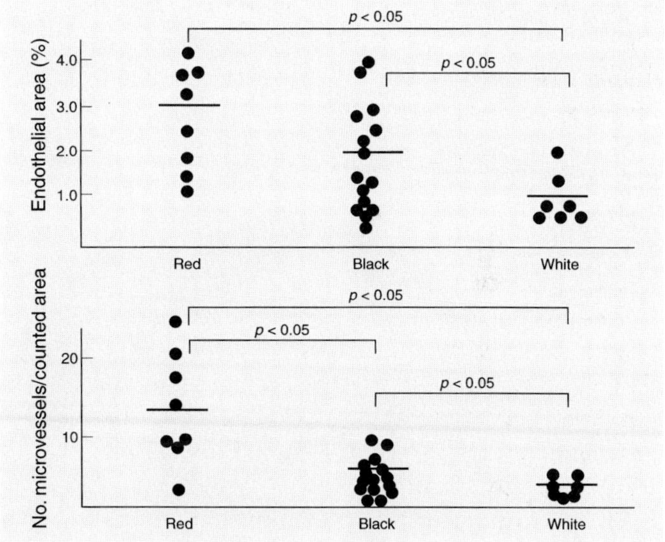

Fig. 4. Upper: Distribution of percentage endothelial area stained by endoglin in each pigmented lesion of endometriosis. Horizontal lines represent the means. Lower: Number of demonstrated microvessels per counted area in different pigmented lesions of endometriosis.

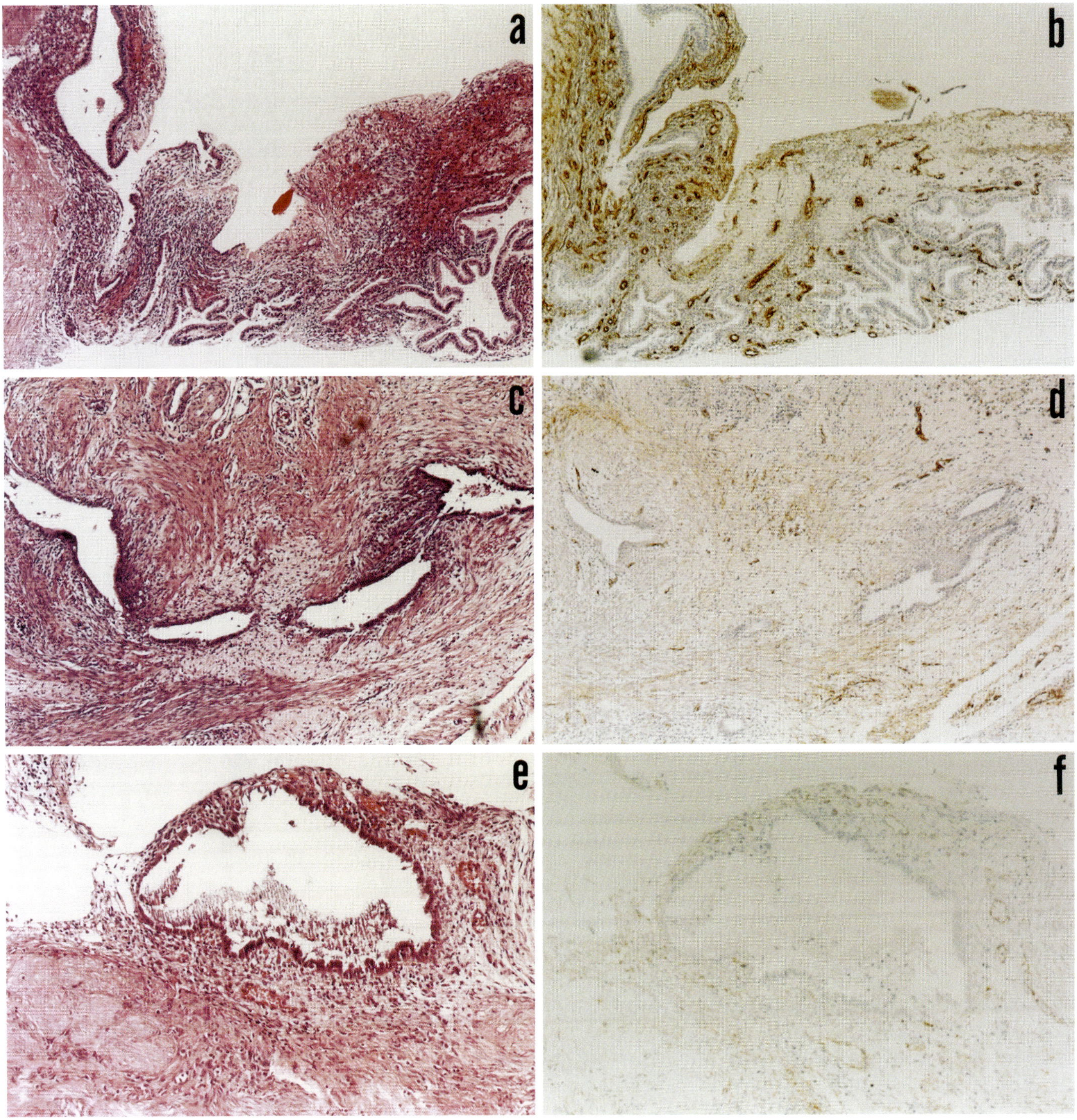

Fig. 5. HE staining of endometriotic tissues taken from different pigmented lesions (**a, c, e**) and immunohistochemical expression of endoglin in the corresponding tissues (**b, d, f**). Tissue specimen of **a** and **b** were taken from red lesions, **c** and **d** from black (blueberry spot) lesions, **e** and **f** from white lesions of pelvic endometriosis. Final magnification was adjusted at ×16 (**a** to **d**) and at ×33 (**e** and **f**).

was significantly increased in red lesions compared to black and white lesions (fig. 4). In fact, the percentage of endothelial area immunostained by endoglin was well correlated with increased vascular formation, and this increased angiogenesis was predominant in red pigmented lesions of endometriosis. The immunohistochemical expression of endoglin in different pigmented lesions is represented in figure 5.

Pathogenesis of Pelvic Endometriosis

The exact pathogenesis and determinants of activity in the natural course of pelvic endometriosis are still unclear, despite the publication of a number of studies. The relationship between ovarian steroid hormones and the development or maintenance of endometriosis has been recognized for decades. A heterogeneity in the expression of estrogen receptors and progesterone receptors in typical endometriosis and eutopic endometrium has been identified, and a variable relationship of these receptors with menstrual cycle has been shown [7, 12]. However, there is no consensus among investigators on the response of endometriosis to ovarian steroid hormones, and it is also not clear whether these changes antecede the development of endometriosis or are results of the disease. Angiogenesis is likely to be involved in the pathogenesis of endometriosis. According to the transplantation theory, when the exfoliated endometrium is attached to the peritoneal layer, the establishment of a new blood supply is essential for the survival of the endometrial implant and development of endometriosis. Cyclic changes are demonstrated in the proliferative capacity and immunohistochemical expression of VEGF and endoglin in eutopic endometrium and different pigmented lesions of endometriosis.

The PCNA index is predominant in the glandular epithelium of eutopic endometrium and progressively increases during the proliferative phase of menstrual cycle. PCNA is a 36-kDa molecule that is widely used to assess the proliferative status of cell populations in a variety of neoplastic conditions and normal tissues. The PCNA index is marked in red lesions compared to black and white lesions in endometriosis. This proliferative capacity in endometriotic tissue was independent of menstrual cycle and cell population. Nisolle et al. [3] also reported high activity in red lesions compared to black lesions using Ki 67 antibody, but cyclic change was recognized in menstrual cycles.

VEGF is a 23-kDa protein that forms a heterodimer and stimulates proliferation of endothelial cells, inducing them to migrate and form tubules. It achieves this by binding to at least two cognate receptors: the fms-like tyrosine kinase (flt) and the kinase-insert domain receptor (KDR) [5]. According to our results, a homogenous expression of VEGF was achieved in glandular epithelium, surrounding stroma and endothelium. In addition, there was no difference in the expression of VEGF among different pigmented lesions of endometriosis and no cyclic variation of VEGF was observed. These findings are in contrast to the report of Donnez et al. [4] who demonstrated that VEGF content was higher in the eutopic glandular epithelium of women with endometriosis during the secretory phase. And significantly higher VEGF content was noted in red lesions as compared with black lesions. The reason of this discrepancy may be due to use of a less sensitive VEGF antibody and to the selection of endometriosis samples. When we measured VEGF concentration in a different population with endometriosis, we found that VEGF concentration was significantly higher in the peritoneal fluid of endometriosis during the proliferative phase of menstrual cycle and this corresponds to the results of McLaren et al. [13]. The computed analysis system for the homogenous identification of VEGF immunostaining should be carefully considered for overestimation in each sample.

Human endoglin (CD105) is a well recognized proliferation-associated antigen (PM) and is expressed at high density on endothelial cells. Up-regulation of endoglin expression has been demonstrated in tumor vasculature and proliferative cells, suggesting that it is a proliferation-associated cell marker [8]. Endoglin has usually been found in endothelial cells of capillaries, arterioles, and venules of a variety of tissues, however this newly identified and highly sensitive angiogenic marker was not examined in endometrium and endometriosis. We showed for the first time immunohistochemical expression of endoglin in eutopic endometrium and different endometriotic lesions. We established that endoglin expression was predominant in stromal tissues of endometrium and endometriosis and no immunolocalization was found in glandular epithelium. This enhanced expression of endoglin is independent of menstrual cycle, although endoglin displays a more prominent expression in the functional layer than in basalis layer at any menstrual phase. In contrast to VEGF, endoglin expression is strongly positive in red lesion comparing to other pigmented lesions of endometriosis.

CD31 (PECAM-1), CD34 and von Willebrand factor have been used as endothelial cell markers, and such factors were stained in relatively large vessels and capillaries. Endoglin is more strongly positive in the endothelial cells of small vessels. The increased cellular expression of endoglin is correlated with increased endothelial surface area occupied by endoglin and also with increased vascular formation. Both the positive area and the number of microvessels is increased in red lesions compared with black and white lesions. This supports the findings of Nissole et al. [12] and Matsuzaki et al. [6], who found more vascular area and increased luminal diameter of vessels in red lesions compared to black lesions. The CSA system for endoglin expression is more sensitive than conventional LSAB method and is suitable for the measurement of the vascular area of immunostained cells and also for the generation of microvessels. These findings suggest that angiogenesis in endometriosis may depend on local regulating factors.

Summary

The pathogenesis and activity of endometriosis possibly depends on internal angiogenesis. The cellular activity of endometriotic lesions does not necessarily coincide with the regulation of angiogenesis. High endoglin levels could provoke an increase in the subperitoneal vessels network and facilitate implantation and viability in the retroperitoneal space. In contrast, lower endoglin levels in black and white lesions could explain the decrease in stromal vascularization followed by inactivation of the implant. Finally, the immunohistochemical study of endoglin expression may be more predictable than VEGF expression for understanding the activity and pathogenesis in the natural course of endometriosis.

References

1 Witz CA, Schenken RS: Pathogenesis. Sem Reprod Endocrin 1997;15:199–208.

2 American Society for Reproductive Medicine: Revised American Society for Reproductive Medicine classification of endometriosis 1996. Fertil Steril 1997;67:817–821.

3 Nisolle M, Casanas-Roux F, Donnez J: Immunohistochemical analysis of proliferative activity and steroid receptor expression in peritoneal and ovarian endometriosis. Fertil Steril 1997;68:912–919.

4 Donnez J, Somes P, Gillerot S, Casanas-Roux F, Nissole M: Vascular endothelial growth factor (VEGF) in endometriosis. Human Reprod 1998;13:1686–1690.

5 Smith SK: Angiogenesis. Sem Reprod Endocrin 1997;15:221–227.

6 Matsuzaki S, Canis M, Darcha C, Dechelotte P, Pouly JL, Bruhat MA: Angiogenesis in endometriosis. Gynecol Obstet Invest 1998;46:111–115.

7 Fujishita A, Nakane PK, Masuzaki H, Chavez RO, Yamabe T, Ishimaru T: Expression of estrogen and progesterone receptors in endometrium and peritoneal endometriosis: an immunohistochemical and in situ hybridization study. Fertil Steril 1997;67:856–864.

8 Burrows FJ, Derbyshire EJ, Tazzari PL, Amlot P, Gazdar AF, King SW, Letarte M, Vitetta ES, Thorpe PE: Up-regulation of endoglin on vascular endothelial cells in human solid tumors: implications for diagnosis and therapy. Clin Cancer Res 1995;1:1623–1634.

9 Gompel C, Silverberg SG (eds): Pathology of Gynecology and Obstetrics: Cyclical Variation of Endometrium. Philadelphia, Lippincott, 1994, pp 171–205.

10 American Society for Reproductive Medicine: Revised American Society for Reproductive Medicine classification of endometriosis 1996. Fertil Steril 1997;67:817–821.

11 Adams JC: Heavy metal intensification of DAB-based HRP reaction product. J Histochem Cytochem 1981;29:775–776.

12 Nisolle M, Menten Y, Casanas-Roux F, Mathieu PE, Wyns C, Donnez J: Immunohistochemical analysis of estrogen and progesterone receptors in endometrium and peritoneal endometriosis. Fertil Steril 1994;62:751–759.

13 McLaren J, Prentice A, Charnock-Jones DS, Smith SK: Vascular endothelial growth factor (VEGF) concentrations are elevated in peritoneal fluid of women with endometriosis. Human Reprod 1996;11:220–223.

Post-Presentation Discussion

Professor Hoshiai: Thank you very much, Dr. Fujishita. Now this interesting paper is open for discussion.

Dr. Suginami: Thank you for a beautiful presentation. You have shown that the remodeling of endometriosis by color and also by expression of PCNA and also the angiogenetic factors. I would like to know about the clear papules of endometriosis. That is, the very early endometriotic change. We can only find, as David Redwine reported, clear papules in the younger generations of the patients like teenagers. And we cannot see any clear papules in the older generation categories of women. So clear papules are representative lesions of early endometriosis. And I would like to know about the PCNA and such in the clear papules, if you have already checked.

Dr. Fujishita: Clear papules in this result are the white lesions, or almost white opacification, or the clear vesicles, which we did not include separately in this study. Because we have taken biopsies of many white lesions, they are histologically reviewed in the endometrial gland and stroma at a very high frequency.

Dr. Suginami: No, I mean clear papules are completely different from white opacified lesions. They are isolated spots, very small spots. Clear papules are often seen in the teenagers or young patients in their 20s, but they are rarely found in the 30s or older patients. They are considered to be very early lesions. Did you check expression of the markers in those very early changes? Those are completely different from the opacified lesions.

Dr. Fujishita: A clear vesicle I understood, and PCNA and VEGF were studied in my materials, but I don't think individual clear papule lesions are included in my materials. But the lesions were included from the epithelium and stroma cells, both of them, but I believe those clear lesions or clear papules may have epithelium, but not stroma cells. We included only seven cases of white lesions. Clear vesicles or serous blebs are not included because they are very few that have the typical epithelial and stroma cells among those clear vesicles and serous blebs. So we did not include clear vesicles or serous bleb lesions. I am very much interested in those lesions and would like to pursue or conduct further studies on these.

Dr. Taketani: The previous two papers pointed to the difference in the intensity of eNOS by Dr. Ota and also the aromatase serum expression by Dr. Kitawaki. Did you find any difference in the intensity in the proliferative activity or the VEGF expression in the endometrium with and without endometriosis?

Dr. Fujishita: It is a very important question, but in this study group, both endometrial curettage and pelvic biopsy were done in only a few cases. I didn't commence to distinguish that relationship.

Dr. Kobayashi: You focused on the VEGF expression in the epithelial cells in both the normal endometrium and endometriosis. As you know, VEGF functions as a paracrine manner because epithelial cells can produce VEGF and then bind to the endothelial cells by a specific receptor for VEGF. Did you find positive staining on the endothelial cells in your study?

Dr. Fujishita: In this study we used epithelial, stromal and also endothelial cells and we found a homogenous expression.

Dr. Kobayashi: The epithelial is also positive for VEGF?

Dr. Fujishita: It is positive for this antibody.

Dr. Kobayashi: Which is intensity staining?

Dr. Fujishita: Endometrium with and without endometriosis? Almost no difference in the immunohistochemical study.

Dr. Kobayashi: I also determined the concentration of VEGF in patients with ovarian cancer and endometriosis by using ascitic fluid. And the result is similar. But the function of VEGF is completely lost, because when we apply Western blot protein analysis of this sample, we can detect very high molecular weight VEGF, which is composed of VEGF and alpha 2 macroglobulin, and which results in the inhibition of VEGF function. VEGF may be involved in the progression of endometriosis. But the VEGF concentration is necessary for the initiation of the disease, but not to maintain the disease status. Do you have any comment?

Dr. Fujishita: That question is very important and the function is very difficult, from my idea. In this study the VEGF concentration is elevated, perhaps due to macrophages, and the VEGF concentration in the proliferative stage is elevated than in the secretory phase. But in the immunostaining there is no difference. I think I cannot answer.

Dr. Kitawaki: Thank you very much for you beautiful presentation. You have mentioned the PCNA and the endoglin immunostaining intensity is different between the different colored lesions. On the other hand, you have mentioned the VEGF concentration is different between patients, not each lesion. So, my question is what are the samples you obtained from the same patients, same lesion, white and red lesions obtained from the same patients, and whether there are any differences between patients, not the lesions.

Dr. Fujishita: In this study, the patient and sampling method was different. I cannot answer your question because tissue expression and PF concentration of VEGF were studied in a separate population.

Dr. Kitawaki: So then, if there are so many more lesions such as white, red and black lesions, what happens in this case?

Dr. Fujishita: I think that in this result, many types that are coexistent in the endometrial pool are not so different in the concentration of VEGF.

Dr. Kitawaki: What happens in the intensity of immunostaining?

Dr. Fujishita: Immunostaining is in some cases of high intensity, but in some lesions it is very low. The expression of immunostaining is very different in the same patient.

Professor Hoshiai: Another question or comment? Because it is very interesting that I found 2 years ago American Society of Reproductive Medicine changed their endometriosis classification by adding the difference of colors of endometrial lesions. Actually, we must take notice of what is different in color that, you know, Dr. Horiday has already reported some activities of differences in colors of lesions, but in detail we do not have so much knowledge, so this is very interesting in this paper.

Dr. Fujimoto: My comment is very similar to Dr. Kobayshi's comment. You know the peritoneal VEGF is not very much changed in confirmation, so my idea is to detect stromal cells' VEGF. Please detect 206 or 189 or 165 or 121 because heavy ice forms with the heparin sulfate. So, binding with the heparin sulfate 206 and 189 works with heparin sulfate as a binding protein. Therefore, it depends on the population, but 165 and 121 is a very directed hit to the KDL, YFDL, FLD. Therefore, if you can organize an isoform, I think you can understand the difference in the endometriosis and the normal endometrium.

Professor Hoshiai: Professor Osteen, do you have any data or could you please give any comment or suggestions? If you have any similar data?

Professor Osteen: I don't believe we have similar data that relate to this question, but again, I think that a lot of what is being looked at in terms of these differences would have to do with the time course of the disease. I still think we are looking for the key factor that would change initially, that would give some indication of a disease process. When I look at your data set again, I think that you are seeing that there are markers that give you better indications of the disease process once it is established. But I still don't think we have a good indication of what are the initial factors that are starting the disease, especially in the younger women. If you look at the kinds of lesions you are seeing in the very, very early stages, what are the key events that we would see at that point? Either with an aromatase in the endometrium perhaps or a particular vascular change that might be causing this conversion. So I would just hope that maybe one of the things we could do is address that particular question looking at the disease in younger women and maybe some of the other uterine markers, maybe together we would get a better indication of where to go. But I don't have data myself that would reflect on what you have seen.

Professor Hoshiai: Thank you very much for your kind addition. Okay. Thank you very much, Dr. Fujishita, for your interesting paper.

Session 3
Review

Gynecologic and Obstetric Investigation

Gynecol Obstet Invest 1999;48(suppl 1):45–56

The Potential Role of Environmental Toxins in the Pathophysiology of Endometriosis

Kaylon L. Bruner-Tran Sherry E. Rier Esther Eisenberg Kevin G. Osteen

Women's Reproductive Health Research Center, Vanderbilt University School of Medicine, Nashville, Tenn., USA

Key Words
TCDD · PCBs · Endometriosis · Progesterone · MMPs · TGF-β

Abstract
Environmental contaminants that are known to disrupt steroid action can influence the development of reproductive diseases. Our group has focused on whether 2,3,7,8-tetrachlorodibenzo-*p*-dioxin (TCDD or dioxin) can disrupt steroid regulation of endometrial matrix metalloproteinase (MMP) expression. The MMPs regulate extracellular matrix turnover in normal tissues, but the inappropriate expression of these enzymes is associated with numerous disease states that involve invasive processes. We have previously shown that secretion of MMPs by human endometrium is critical for establishment of ectopic lesions in a nude mouse model of experimental endometriosis. In this report, we show that TCDD exposure promotes establishment of experimental endometriosis by interfering with the ability of progesterone to suppress endometrial MMP expression.

Introduction

Although endometriosis, the presence of endometrial glands and stroma at an extrauterine site, was first described more than a century ago [1], the precise cause of this condition remains unknown. The most widely accepted theory for the development of endometriosis suggests that endometrial fragments, sloughed during menstruation, could be deposited, via the fallopian tubes, into the peritoneal cavity [2]. Additional theories have been advanced to explain the development of endometriosis in unusual sites [3] or, more rarely, in men [4]. Nevertheless, clinical evidence and basic research studies support Sampson's metastatic theory as the primary mechanism by which endometriosis is established. The survival of endometrial tissue in the peritoneal cavity may be related to the local steroid environment [5], including the ability of tissue fragments to implant at ectopic sites [6].

Establishment of endometriosis is an invasive process that requires degradation of the extracellular matrix [6]. A number of matrix metalloproteinases (MMPs) are expressed in the endometrium in association with cellular proliferation and during menstrual tissue breakdown [7, 8]. Secretion of MMPs by endometrial tissue fragments could mediate peritoneal invasion, required for establishment of ectopic sites of growth [reviewed by 9]. For example, estradiol promotes MMP expression during cellular proliferation in the normal endometrium and this steroid

KARGER
Fax + 41 61 306 12 34
E-Mail karger@karger.ch
www.karger.com

0378–7346/99/0485–0045$17.50/0

Accessible online at:
www.karger.com/journals/goi

Kevin G. Osteen, PhD
Vanderbilt University School of Medicine, MCN B-1100
1161 21st Ave South, Nashville TN 37232 (USA)
Tel. +1 615 322 4196, Fax +1 615 343 7913
E-Mail kevin.osteen@mcmail.vanderbilt.edu

promotes similar growth at ectopic sites of endometriosis [10]. In addition, aromatase inhibitors, which reduce production of estradiol, have recently been used to successfully treat an aggressive form of endometriosis [11]. In contrast to estrogen effects, progesterone is a potent suppressor of endometrial MMPs during endometrial differentiation [10, 12, 13]. Clinically, progestins have been used to treat endometriosis [14] and the progesterone-dominated environment of pregnancy can regress endometriotic lesions in some women [10]. Experimentally, exposure of human endometrial fragments to progesterone blocks MMP expression and prevents the development of experimental endometriosis in a nude mouse model [15].

Given the pathophysiology of this disease, it is reasonable to assume that the development of endometriosis may be impacted by environmental chemicals that disrupt steroid action. Environmental toxins such as TCDD are of particular concern to the disease process of endometriosis, because this agent can disrupt the synthesis of ovarian steroids [16, 17] as well as cytokines, which are important secondary mediators of steroid action [18], including MMP regulation [19].

The expression of MMPs appears inappropriate or elevated in ectopic lesions removed from women with endometriosis [9, 20, 21]. In addition, Wegner et al. [22] demonstrated that MMP-3 protein may be continuously expressed in the eutopic endometrium of women with endometriosis, and we have found a similar pattern of expression of MMP-11 mRNA [Osteen, unpublished observation, 1999]. Interestingly, progesterone exposure is associated with an increase in endometrial cell apoptosis [23], and both eutopic and ectopic endometrium of women with endometriosis show a reduction in apoptosis [23, 24]. Progesterone responsiveness in endometriotic lesions appears to be diminished [25, 26] and, taken together, the above studies suggest that progesterone sensitivity may also be reduced in eutopic endometrium of women with the disease.

Unfortunately, the cellular mechanisms by which endometrial tissue or ectopic sites of endometriosis becomes less sensitive to progesterone action are not clear. It is possible that 2,3,7,8-tetrachlorodibenzo-*p*-dioxin (TCDD or dioxin) and dioxin-like compounds may influence the development of endometriosis by diminishing endometrial response to progesterone or by disrupting progesterone-mediated cell-cell communication. For example, a number of cytokines, which are associated with progesterone action in the endometrium, are disrupted by TCDD in other systems [18, 27]. Because the regulation of endometrial MMPs by progesterone requires cytokine mediators of cell-cell communication [12, 19], our laboratory has explored the role TCDD might play in disrupting the expression of these enzymes. In the studies described below we will demonstrate the stimulatory impact of human endometrial TCDD exposure in vitro on the establishment of ectopic lesions in experimental endometriosis. In addition, we will show that TCDD can specifically interfere with cell- cell communication by disrupting the regulation of TGF-β_2, a key factor in endometrial MMP regulation.

Environmental Endocrine Disruptors and Endometriosis

The prevalence of endometriosis appears to be on the rise and this increase can only be partially explained by better physician awareness and diagnosis [28]. It has been postulated that development of certain diseases, including endometriosis, may be related to the industrial contamination of our environment [29, 30]. Our laboratory has focused on polychlorinated diaromatic hydrocarbons (PCDHs), as these agents are known to have endocrine disrupting properties that may specifically interfere with steroid action. Of these substances, TCDD is considered to be the most potent and is used as a reference compound for the effects of all other PCDHs. Although TCDD is an unintentional by-product of manufacturing, polychlorinated biphenyls (PCBs) were specifically developed for use in a number of industrial processes, resulting in worldwide environmental contamination. These compounds degrade very slowly in the environment and, due to their lipophilic nature, tend to bioaccumulate and biomagnify within the food chain [for review see 31]. In the general population, it is believed that ingestion of contaminated foods is the most likely source of TCDD and other PCDHs [32], but accidental exposure has also occurred in the workplace, during wartime, and as a result of numerous industrial accidents [31, 33, 34].

The effects of the PCDHs appear to be primarily mediated via the aryl hydrocarbon receptor (AhR). The AhR is an orphan nuclear receptor whose natural ligand is not currently known, but binds TCDD and other dioxin-like compounds with very high affinity. Binding of ligand results in activation and translocation of the AhR complex to the nucleus where DNA binding and transcriptional regulation occurs [reviewed by 35]. TCDD has been shown to disrupt the expression of numerous proteins [reviewed by 18] and a number of animal studies have indicated a wide range of adverse effects, which include teratogenicity, increases in cancer, and thymic involution [reviews by 36, 37]. Similar effects on human populations are beginning to be appreciated and it is becoming clear that TCDD can be a potent carcinogen, immunosuppressive agent, and reproductive toxin [29, 33, 37, 38].

Although data regarding adverse effects of PCDHs on reproduction are limited, the rate of spontaneous abortion is increased in women and primates exposed to TCDD prior to or during pregnancy [39, 40]. There is also compelling experimental evidence of an effect of TCDD on the development of endometriosis. Rier et al. [41] published a retrospective analysis of endometriosis occurring spontaneously in macaque monkeys exposed to dietary TCDD for 5 years. Seven years after discontinuation of the study (and TCDD administration), one animal died as a result of severe infiltrating endometriosis. Two years later, a second monkey, also in the group exposed to the highest levels of TCDD, died due to endometriosis-associated bowel obstruction. After the death of the second monkey, all female animals underwent laparoscopy and moderate to severe endometriosis was detected in 71 and 86% of monkeys treated with 5 and 25 ppt TCDD, respectively. Only 33% of control animals (not exposed to TCDD) showed evidence of minimal endometriosis. In light of recent observations indicating extensive contamination of the food chain with a variety of PCDHs [32], this same primate population has recently been reevaluated with regard to serum levels of TCDD and PCBs. Surprisingly, serum concentrations of TCDD did not correlate with the presence of endometriosis. However, a strong correlation was found between exposure to TCDD, elevated serum concentrations of PCBs, and symptomatic endometriosis [42]. The authors suggest that TCDD exposure may result in alterations of normal physiologic processes that can enhance effects of other environmental toxins. Therefore, prior exposure to TCDD, rather than actual body burden, may be the most critical component associated with the development of endometriosis in these animals.

These and other studies in experimental models indicate a link exists between TCDD exposure and the development of endometriosis in the laboratory [41–43]. In women, a direct correlation between endometriosis and environmental endocrine disruptors has not been firmly established, but evidence is beginning to accumulate. For example, in 1992, Gerhard et al. [44] found higher levels of PCB congeners in infertile women with endometriosis than in women without the disease. In Belgium, TCDD concentrations in breast milk is among the highest in the world [45] and the incidence of endometriosis is 60–80% among women treated for infertility [46].

Table 1. Effects of TCDD on the development of endometriotic-like lesions in nude mice

Number of mice	Tissue treatment	Mouse treatment	Total lesions
8	E	E	20
8	EP	E	0
8	E + TCDD	E	42
8	EP + TCDD	E	48

All mice were implanted with an estradiol-releasing pellet 24 h prior to intraperitoneal injection of human endometrial tissues. Mice were intraperitoneally injected with proliferative phase endometrial tissues treated in vitro for 24 h with estradiol (E = 10^{-8} *M*), estradiol plus progesterone (EP, E = 10^{-9} *M*; P = 5 × 10^{-7} *M*), E + 2,3,7,8-tetrachlorodibenzo-*p*-dioxin (TCDD; 10^{-6} *M*) or EP + TCDD. Ten days after tissue injection, mice were killed and necropsied for endometriotic-like disease.

Despite the compelling epidemiology and basic research data noted here, little is known regarding the basic cellular mechanisms by which TCDD may actually influence the development of endometriosis. To initially address this question, we utilized an established nude mouse model of endometriosis that mimics many aspects of the human disease. In our model, normal proliferative phase human endometrium is obtained by biopsy and can be exposed to steroids or toxins in vitro. Earlier studies showed that treatment of endometrial organ cultures with estradiol maintains expression of MMPs whereas progesterone treatment suppresses these enzymes [12, 13].

After in vitro treatments, endometrial tissues are intraperitoneally injected into ovariectomized nude mice with subcutaneous estradiol-releasing pellets. Within 5–10 days of tissue injection, 95% of mice receiving estradiol-treated tissue develop lesions, although progesterone treatments prevent the experimental disease [15, 47]. Regulation of MMPs appears to be the primary mechanism by which steroids promote or prevent experimental endometriosis, because coinjection of TIMP-1 (which blocks MMP action) dramatically reduces lesion formation regardless of estradiol treatment [15].

To investigate the impact of TCDD exposure in this experimental model, we cotreated endometrial samples within each steroid group with TCDD prior to intraperitoneal injection into nude mice. As shown in table 1, we

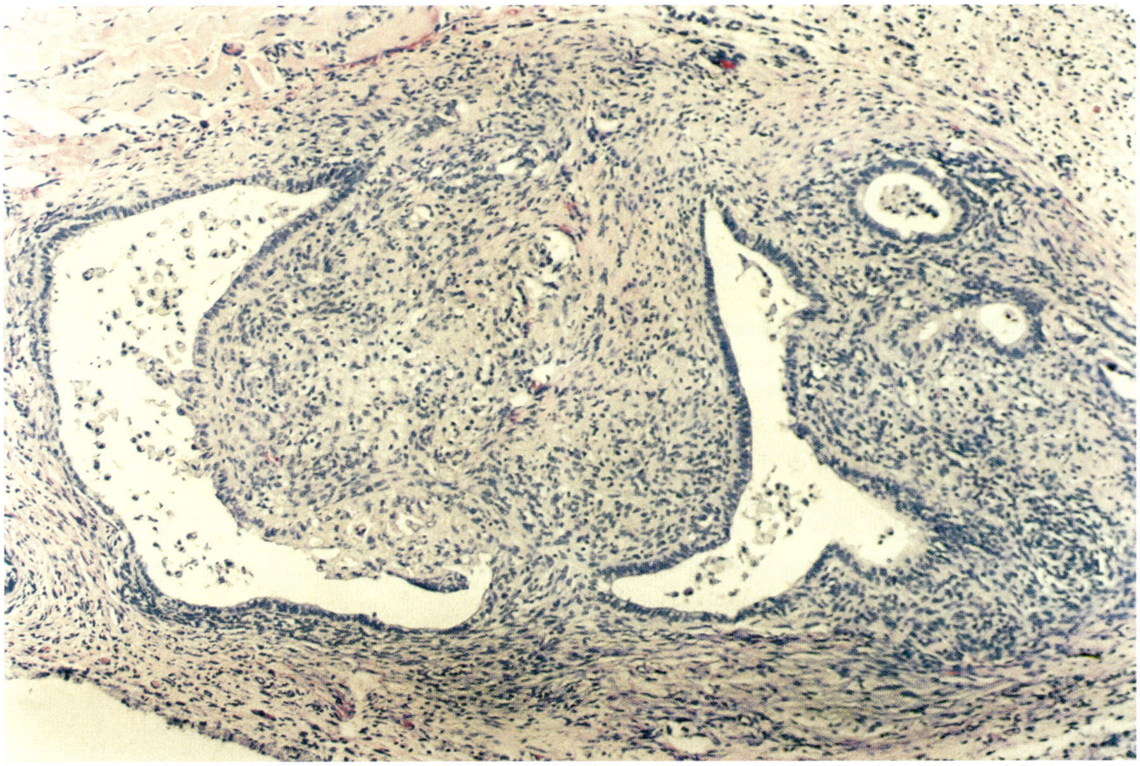

1A

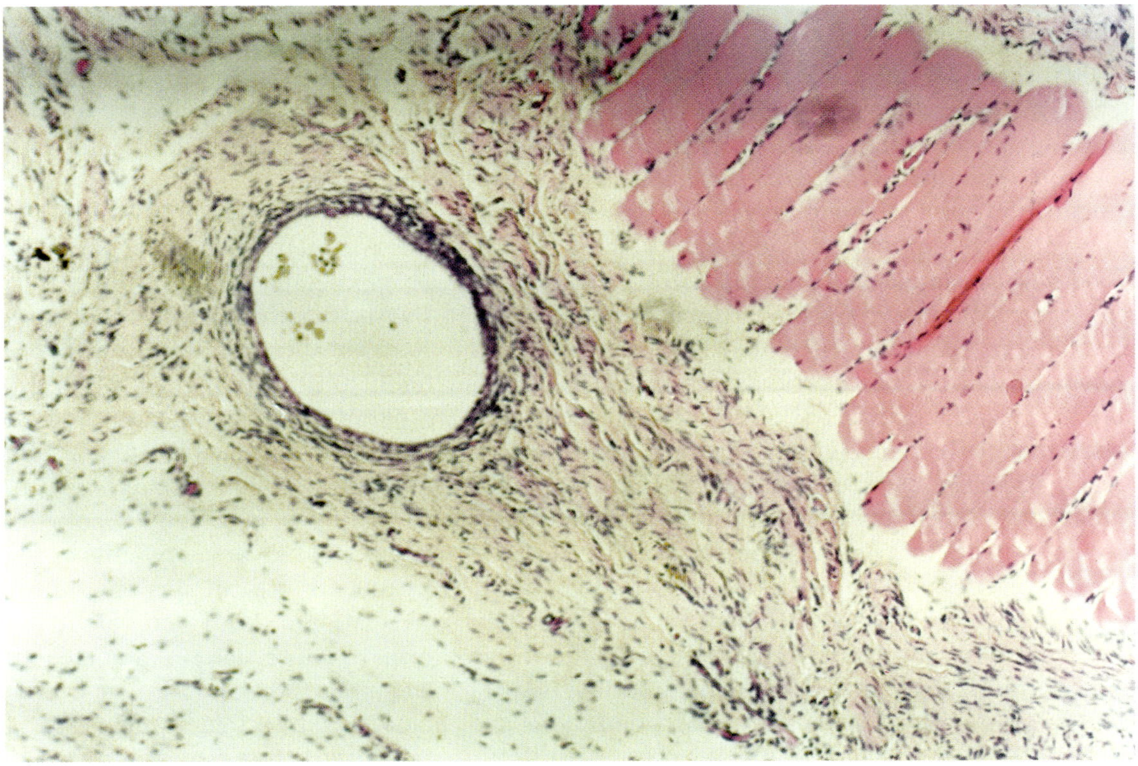

1B

found that TCDD treatment increased the number of peritoneal lesions in mice twofold over treatments with estradiol alone. More importantly, TCDD treatments blocked the ability of progesterone to prevent ectopic human tissue lesions in mice. In addition, individual lesions removed from nude mice receiving TCDD-treated tissues were consistently larger (fig. 1A) than individual lesions removed from mice receiving tissues treated with estradiol alone (fig. 1B).

Our observations are similar to findings reported by other investigators who have described a more aggressive disease after exposure to TCDD [41, 43]. Numerous studies have indicated that TCDD can interfere with normal estrogen action [48, 49], including prevention of estradiol-associated increases in the progesterone receptor [49]. However, our studies are among the first to provide evidence that TCDD is able to interfere more directly with progesterone action relative to a specific process required for establishment of endometriosis.

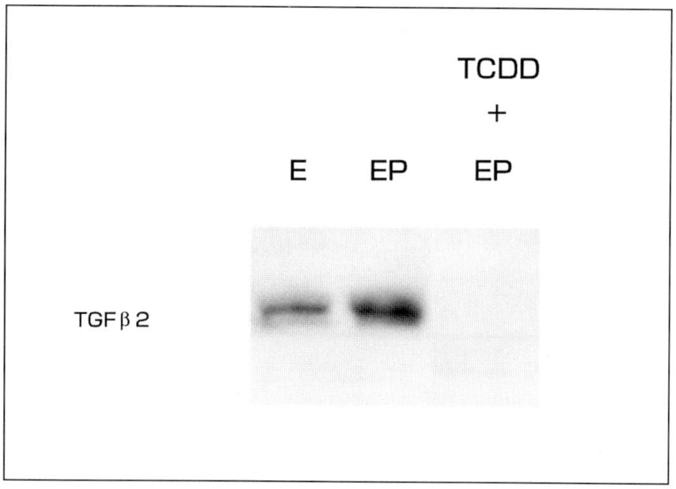

Fig. 2. Western blot analysis of expression and regulation of TGF-β_2 in endometrial organ cultures. Proliferative phase endometrium was established in organ culture as previously described [15] and treated for 48 h with estradiol (E, 10^{-8} *M*), estradiol plus progesterone (EP, E = 10^{-9} *M*; P = 5×10^{-7} *M*) or estradiol, progesterone and 2,3,7,8-tetrachlorodibenzo-*p*-dioxin (TCDD) (EP + TCDD; E = 10^{-9} *M*; P = 5×10^{-7} *M*; TCDD = 10^{-6} *M*). Western blot analysis was performed as previously described [56].

TCDD as a Disruptor of Progesterone Action

Although epidemiologic and experimental data suggest a role for toxins such as TCDD in the pathophysiology of endometriosis, few studies have directly addressed a specific cellular mechanism by which TCDD may disrupt steroid action during the development of this disease. In most systems, TCDD is known to have anti-estrogen action, including disruption of estrogen associated protein expression [50], increased estrogen metabolism [51, 52] and decreased estrogen receptor expression [53]. Because much evidence exists indicating that endometriosis is an estrogen-dependent disease [reviewed by 10], it is unclear why TCDD, as an antiestrogen, would promote development of this disease. Clinical and experimental evidence indicates an important role for progestins in treating endometriosis [10, 14], and TCDD can disrupt ovarian synthesis of progesterone [16]. Moreover, in cultured keratinocytes, TCDD has been shown to inhibit the differentiation-associated expression of TGF-β_2 [27], a growth factor that mediates progesterone-associated suppression of endometrial MMPs [19].

In our studies, the secondary action of TGF-β_2 appears essential for suppression of endometrial MMP expression [19], an action of progesterone clearly associated with the prevention of experimental endometriosis [54]. Not surprisingly, progesterone-induced expression of TGF-β_2 is also critical to maintain an appropriate endometrial environment during pregnancy. For example, a subset of women with recurrent spontaneous abortion were found to be deficient in decidua-associated TGF-β_2 expression [55]. Interestingly, human studies and studies in nonhuman primates indicate that TCDD exposure increases spontaneous abortion [39, 40], a uterine event that may be linked mechanistically to the development of endometriosis [41, 46].

Because pregnancy or prevention of experimental endometriosis appears to specifically require local TGF-β_2 expression, we explored the effect of TCDD on regulation of TGF-β_2 protein in the normal endometrium. Proliferative phase endometrial tissues were established as organ cultures and treated for 48 h with E (estradiol), E plus P (progesterone) or E, or P and TCDD. As shown in figure 2, TCDD exposure completely blocks progesterone-

Fig. 1. Hematoxylin and eosin staining of human endometrium growing in a nude mouse. Proliferative phase human endometrial tissues were treated in vitro for 24 h with estradiol (E = 10^{-8} *M*) or estradiol plus 2,3,7,8-tetrachlorodibenzo-*p*-dioxin (TCDD) (E = 10^{-8} *M*; TCDD = 10^{-6} *M*) prior to be injected into ovariectomized nude mice previously implanted with an estradiol-releasing pellet (1.5 mg/60-day release). Endometriotic-like lesions developing from estradiol + TCDD-treated tissues (**A**) are larger with a more pronounced stromal compartment than lesions established from (**B**) tissues treated only with estradiol. Magnification 200 ×.

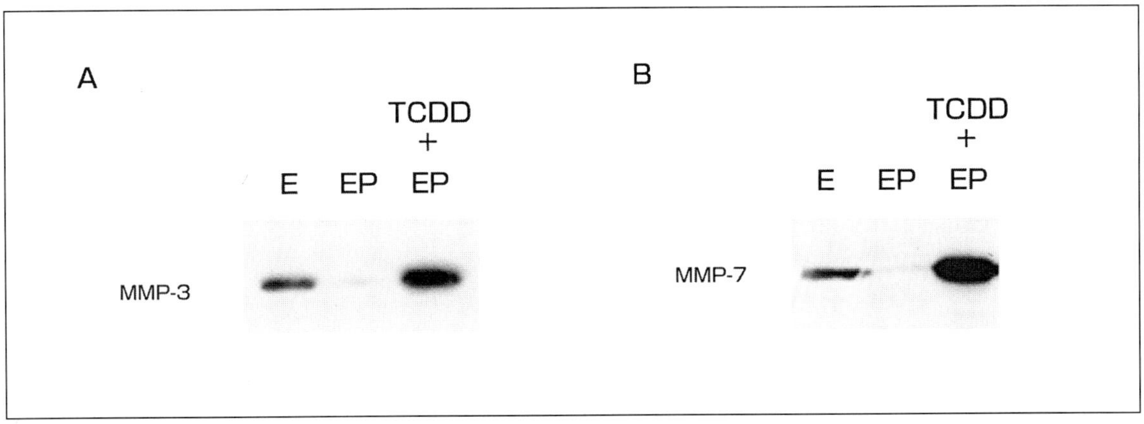

Fig. 3. Western blot analysis of (**A**) matrix metalloproteinase-3 (MMP-3) and (**B**) MMP-7 expression and regulation in endometrial organ cultures. Proliferative phase endometrium was established in organ culture as previously described [15] and treated for 48 h with estradiol (E, 10^{-8} *M*), estradiol plus progesterone (EP, E = 10^{-9} *M*; P = 5 × 10^{-7} *M*) or estradiol, progesterone and 2,3,7,8-tetrachlorodibenzo-*p*-dioxin (TCDD) (EP + TCDD; E = 10^{-9} *M*; P = 5 × 10^{-7} *M*; TCDD = 10^{-6} *M*).

mediated induction of endometrial TGF-β_2 expression. Not surprisingly, TCDD interference of induction of TGF-β_2 was also associated with a failure to suppress MMP-3 and MMP-7 (fig. 3). The ability of TCDD to block progesterone associated TGF-β_2 expression provides an important clue to understanding how this environmental endocrine disruptor may lead to the development of endometriosis in women.

Summary

In concert with the benefits of industrialization of our society, our environment has become contaminated with a wide variety of chemical pollutants. Many of these contaminants are known to be endocrine disruptors, but their true effects alone or in combination have yet to be fully assessed. We have focused our efforts on understanding what role, if any, TCDD may have in the development of endometriosis in an experimental model. Our studies have shown that TCDD, an environmental contaminant with antiestrogenic effects, can also disrupt progesterone action. More specifically, we demonstrated that exposure of normal endometrial tissues to TCDD prevents the progesterone-associated increase in TGF-β_2. Because TGF-β_2 is critical to maintaining normal pregnancy and preventing maternal rejection of the fetus [55, 56] interfering with the production of this growth factor may have broad implications to reproductive processes other than endometriosis. However, this growth factor plays a central role in progesterone-mediated suppression of MMPs, including the ability of progesterone to prevent experimental endometriosis. Progestins are often used as therapeutic agents for women with endometriosis [14] and the high progesterone levels associated with pregnancy appear to protect women from the development of this disease [10].

The ability of TCDD to modify or interfere with progesterone action may begin to explain numerous observations that have suggested an association of TCDD with endometriosis. Clearly, we find that TCDD exposure can block the ability of progesterone to prevent experimental endometriosis, an effect correlated with this toxin's ability to inhibit progesterone-associated TGF-β_2 expression and MMP suppression. Nevertheless, a better understanding of progesterone action and the broader role TCDD may play in disruption of steroid action is needed. In particular, recent studies suggesting that PCBs may act in concert with an initial exposure to TCDD or other dioxin-like compounds [42] deserve closer examination. Humans and animals are exposed to a complex mixture of environmental toxins and future studies must begin to examine TCDD and its congeners relative to additive or synergistic effects. Recent advances in the ability to quantify a variety of PCDHs in tissues and body fluids will allow closer examination of exposed human and animal populations, providing important clues to understanding endocrine disrupting chemicals and disease.

References

1 Von Rokitansky C: Über Uterusdrüsen-Neubildung im Uterus und Ovarialsarcomen. Zkk Gesellsch d Aerzte zu Wien 1860;37:577.

2 Sampson JA: Peritoneal endometriosis due to menstrual dissemination of endometrial tissue into the peritoneal cavity. Am J Obstet Gynecol 1927;14:422–469.

3 Suginami H: A reappraisal of the celomic metaplasia theory by reviewing endometriosis occurring in unusual sites and instances. Am J Obstet Gynecol 1991;165:214–218.

4 Fujii S: Secondary Mullerian system and endometriosis. Am J Obstet Gynecol 1991;6: 121–127.

5 DiZerga GS, Barber DL, and Hodgen GD: Endometriosis: Role of steroids in initiation, development, maintenance and suppression. Fertil Steril 1980;33:649–653.

6 Spuijbroek MDEH, Dunselman GAJ, Menheere PPCA, Evers JLH: Early endometriosis invades the extracellular matrix. Fertil Steril 1992;58:929–933.

7 Rodgers WH, Osteen KG, Matrisian LM, Navre M, Gorstein F: Expression and localization of matrilysin, a matrix metalloproteinase, in the human endometrium. Am J Obstet Gynecol 1993;168:253–260.

8 Rodgers WH, Matrisian LM, Giudice LC, Dsupin B, Cannon P, Svitek C, Gorstein F, Osteen KG: Patterns of matrix metalloproteinase expression in cycling endometrium imply differential functions and regulation by steroids. J Clin Invest 1994;94:946–953.

9 Osteen KG, Bruner KL, Sharpe-Timms, KL: Steroid and growth factor regulation of matrix metalloproteinase expression and endometriosis. Semin Reprod Endocrinol 1996;14:247–255.

10 Halme J, Stovall D: Endometriosis and its medical management; in Wallach EE and Zacur HA (eds): Reproductive Medicine and Surgery. St. Louis, Mosby, 1995, pp 695–710.

11 Takayama K, Zeitoun K, Gunby RT, Sasano H, Carr BR, Bulun SE: Treatment of severe post-menopausal endometriosis with an aromatase inhibitor. Fertil Steril 1998;69:709–713.

12 Osteen KG, Rodgers WH, Gaire M, Hargrove J, Gorstein F, Matrisian: Stromal-epithelial interaction mediates steroidal regulation of matrix metalloproteinase expression in the human endometrium. Proc Natl Acad Sci USA 1994; 91:10129–10133.

13 Schatz F, Papp C, Toth-pal E, Lockwood C: Ovarian-steroid modulation of stromelysin-1 expression in human endometrial stroma and decidual cells. J Clin Endocrinol Metab 1994; 78:1467–1472.

14 Olive DL: Medical treatment: Alternatives to danazol; in Schenken RS, (ed): Endometriosis: Contemporary Concepts and Clinical Management. Philadelphia, Lippincott, 1989, pp 192–202.

15 Bruner KL, Matrisian, LM, Rodgers WH, Gorstein F, Osteen KG: Suppression of matrix metalloproteinases inhibits establishment of ectopic lesions by human tissue in nude mice. J Clin Invest 1997;99:2851–2857.

16 Enan E, Lasley B, Stewart D, Overstreet J, Vandervoort CA: 2,3,7,8-tetrachlorodibenzo-*p*-dioxin (TCDD) modulates function of human luetinizing granulosa cells via cAMP signalling and early reduction of glucose transporting activity. Reprod Toxicol 1996;10:191–198.

17 Heimler I, Rawlins RG, Owen H, Hutz RJ: Dioxin perturbs in a dose and time-dependent fashion, steroid secretion, and induces apoptosis of human luteinized granulosa cells. Endocrinology. 1998;139:4373–4379.

18 Dohr O, Vogel C, Abel J: Effects of 2,3,7,8-tetrachlorodibenzo-*p*-dioxin (TCDD) on growth factor expression in the human breast cancer cell line MCF-7. Arch Toxicol 1995;69: 259–265.

19 Bruner KL, Rodgers WH, Korc M, Gold L, Hargrove J, Matrisian LM, Osteen KG: Transforming growth factor-β mediates the progesterone suppression of an epithelial metalloproteinase by adjacent stroma. Proc Natl Acad Sci USA 1995;92:7362–7366.

20 Saito T, Mizumoto H, Kuroki K, Fujii M, Mori S, Kudo R: Expression of MMP-3 and TIMP-1 in endometriosis and the influence of Danazol. Acta Obstet Gynecol Japan 1995;47:495–496.

21 Sharpe-Timms KL, Keisler LW, McIntush EW, Keisler DH: Tissue inhibitor of metalloproteinase-1 concentrations are attenuated in peritoneal fluid and sera of women with endometriosis and restored in sera by gonadotropin releasing hormone-agonist therapy. Fertil Steril 1998;69:1128–1134.

22 Wegner EL, McBean JH, Khavari A, Brumsted JR: Abnormal MMP-3 expression in the endometrium of women with endometriosis. Presented at the 52nd Annual Meeting of the American Society for Reproductive Medicine (abstr.). Boston, 1996.

23 Jones RK, Searle RF, Bulmer JN: Apoptosis and BCL-2 expression in normal human endometriosis, endometriosis and adenomyosis. Hum Reprod 1998;13:3496–3502.

24 Dmowski WP, Gebel H, Braun DP: Decreased apoptosis and sensitivity to macrophage mediated cytolysis of endometrial cells in endometriosis. Hum Reprod 1998;4:696–701.

25 Lessey BA, Metzger DA, Haney AF, McCarty KS Jr: Immunohistochemical analysis of estrogen and progesterone receptors in endometriosis: Comparison with normal endometrium during the menstrual cycle and the effect of medical therapy. Fertil Steril 1989;49:229–235.

26 Bergqvist A, Ferno M: Oestrogen and progesterone receptors in endometriotic tissue and endometrium: Comparison of different cycle phases and ages. Hum Reprod 1993;8:2211–2217.

27 Gaido KW, Maness SC, Greenlee WF: 2,3,7,8-tetrachlorodibenzo-*p*-dioxin (TCDD) regulation of TGF-a and b2 expression in a human kertinocyte cell line involves both transcriptional and post-transcriptional control. J Biol Chem 1992;267:24591–24595.

28 Molpus KL: Pathophysiology and management of prior or current endometriosis in the perimenopausal and postmenopausal woman; in Eisenberg E (ed): Infertility and Reproductive Clinics of North America. Philadelphia, Saunders, 1995, pp 805–828.

29 Osteen KG, Sierra-Rivera E: Does disruption of immune and endocrine systems by environmental toxins contribute to the development of endometriosis? Semin Reprod Endocrinol 1997;15:301–308.

30 Watanabe S, Kitamura K, Nagahashi M: Effects of dioxins on human health: A review. J Epidemiol 1999;1:1–13.

31 Safe SH: Polychlorinated biphenyls (PCBs): Environmental impact, biochemical and toxic responses, and implications for risk assessment. Crit Rev Toxicol 1994;24:87–149.

32 Sweeney A: Reproductive epidemiology of dioxins; in Schecter A (ed): Dioxins and Health, New York, Plenum Press, 1994, pp 549–583.

33 Holsapple M, Snyder N, Wood S, Morris D: A review of 2,3,7,8-tetrachlorodibenzo-*p*-dioxin (TCDD) induced changes in immunocompetence. Toxicology 1991;69:219–255.

34 Bertazzi PA, Pesatori AC, Consonni D, Tironi A, Landi MT, Zoccheti C: Cancer incidence in a population accidentally exposed to 2,3,7,8-tetrachlorodibenzo-*p*-dioxin (TCDD). Epidemiology 1993;4:398–406.

35 Greenlee WF, Sutter TR, Marcus C: Molecular basis of dioxin actions on rodent and human target tissues. Prog Clin Biol Res 1994;387:47–57.

36 Poland A Knutson JC: 2,3,7,8-tetrachlorodibenzo-*p*-dioxin and related halogenated aromatic hydrocarbons: Examination of the mechanism of toxicity. Annu Rev Pharmacol Toxicol 1992;22:517–554.

37 Huff J, Lucier G, Tritscher A: Carcinogenicity of TCDD: Experimental, mechanistic and epidemiologic evidence. Annu Rev Pharm 1994; 34:343–372.

38 Birnbaum LS: The development effects of dioxins. Environ Health Perspect 1995;103: 89–94.

39 Damewood M: The association of endometriosis and repetitive (early) spontaneous abortions. Semin Reprod Endocrinol 1989;7:155–160.

40 Bowman RE, Schantz SL, Weerasinghe NCA, Gross M, Barsotti D: Chronic dietary intake of 2,3,7,8-tetrachlorodibenzo-*p*-dioxin (TCDD) at 5 or 25 ppt in the monkey: TCDD kinetics and dose-effect estimates of reproductive toxicity. Chemosphere 1989;18:243–252.

41 Rier SE, Martin DC, Bowman RE, Dmowski WP, Becker JL: Endometriosis in rhesus monkeys *(Macaca mulatta)* following chronic exposure to 2,3,7,8-tetrachlorodibenzo-*p*-dioxin. Fundam Appl Toxicol 1993;21:433–441.
42 Rier SE, Turner WE, Martin DC, Morris R, Lucier GW, Clark GC: Serum levels of TCDD and dioxinlike chemicals in rhesus monkeys chronically exposed to TCDD: Correlation of increased serum PCB levels with endometriosis. Environ Health Perspect 1999 (in press).
43 Cummings AM, Metcalf JL, Birnbaum L: Promotion of endometriosis by 2, 3, 7, 8-tetrachlorodibenzo-*p*-dioxin in rats and mice: Time-dose dependence and species comparison. Toxicol Appl Pharmacol 1996;138:131–139.
44 Gerhard I, Runnebaum B: Limits of hormone substitution in the case of exposure to harmful substances and fertility disturbances. Zentralbl Gynäkol 1992;114:593–602.
45 WHO Report: Levels of PCBs, PCDDs and PCDFs in breast milk: Result of WHO coordinated interlaboratory quality control studies and analytical field studies. WHO Environmental Series, 1989.
46 Koninckx PR, Braet P, Kennedy SH, Barlow DH: Dioxin pollution and endometriosis in Belgium. Hum Reprod 1994;9:1001–1002.
47 Bruner KL, Keller NR, Osteen KG: Interleukin-1α opposes suppression of human endometrial matix metalloproteinases by progesterone in a model of experimental endometriosis. Proceedings of the VI World Congress on Endometriosis, 1998.
48 Safe S, Astroff B, Harris M, Zacharewski T, Dickerson R, Romkes M, Biegel L: TCDD and related compounds as anti-estrogens: Characterization and mechanism of action. Pharmacol Toxicol 1991;69:400–409.
49 Safe S, Wang F, Porter W, Duan R, McDougal A: Ah receptor agonists as endocrine disruptors: Antiestrogenic activity and mechanisms. Toxicol Lett 1998;102–103:343–347.
50 Wang W, Smith R 3rd, Safe S: Aryl hydrocarbon receptor-mediated anti-estrogenicity in MCF-7 cells: Modulation of hormone induced cell cycle enzymes. Arch Biochem Biophys 1998;356:239–248.
51 Gierthy JF, Lincoln DW, Gillespie MB, Seeger JI, Martinez HL, Dickerman HW, Kumar SA: Suppression of estrogen regulated extracellular tissue plasminogen activator activity of MCF-7 cells by 2,3,7,8-tetrachlordibenzo-*p*-dioxin. Cancer Res 1987;47:6198–6203.
52 Gierthy JF, Linclon DW 2d, Kampcik SJ, Dickerman HW, Bradlow HL, Niwa T, Swaneck GE: Enhancement of 2- and 16 alpha-estradiol hydroxylation in MCF-7 human breast cancer cells by 2,3,7,8-tetrachlordibenzo-*p*-dioxin. Biochem Biophys Res Commun 1988;157:515–520.
53 Tian Y, Ke S, Thomas T, Meeker RJ, Gallo MA: Transcriptional regulation of estrogen receptor gene expression by TCDD. J Steroid Biochem Mol Biol 1998;67:17–34.
54 Bruner KL, Eisenberg E, Gorstein F, Osteen KG: Progesterone and transforming growth factor-β coordinately regulate suppression of endometrial matrix metalloproteinases and prevention of experimental endometriosis. Steroids 1999;64:648–653.
55 Lea RG, Underwood RJ, Flanders KC, Hirte H, Branwatt D, Michel M, Daya S, Harley C, Mowbray JF, Clark DA: A subset of patients with recurrent spontaneous abortion is deficient in TGF-β_2-producing suppressor cells in decidua near the placental attachment site. Am J Reprod Immun 1994;35:52–64.
56 Clark DA, Vince G, Flanders KC, Kivte H, Starkey P: CD56+ lymphocytic cells in human first trimester decidua as a source of novel TGF-β_2-related immunosuppressive factors. Hum Reprod 1994;9:22701–2277.

Post-Presentation Discussion

Professor Osteen: Again, I want to thank the organizers for the invitation and what has been a very stimulating session. It is very nice to see familiar faces and get a sense of the endometriosis community in Japan. I think that the investigators who are here collectively are an important force in this field and I think it is nice to get a sense of that by coming to Japan and I really appreciate this opportunity.

Professor Taketani: I think Professor Osteen clearly demonstrated the effect of the activity of MMP in determining the endometrial tissue implanted in the nude mouse. It is very exciting data. Now his talk is open for discussion. I think we are going to have many questions to add to his speech. Any questions? Both English and Japanese are acceptable.

Dr. Tatsumi: Thank you for your presentation. I am also concerned about dioxin and endometriosis. I think endometriosis is the most sensitive human disorder. I think it may be concerned with dioxin. In your presentation, MMP in vitro was affected by TCDD. What is the concentration of TCDD in nude mouse? What dose of TCDD? I am very interested in this.

Professor Osteen: The experiments to date that we have done, have involved taking a biopsy of the human endometrium and exposing it to doses of dioxin in vitro and then injecting into nude mice which have not been exposed to TCDD. So, what we find is that the prior exposure of the tissue impacts the development of the endometrial tissue in the nude mouse. We have yet to report on the studies we are doing now to determine whether or not feeding the animals dioxin would have an impact. One of the reasons of that is, as you know, working with dioxin is not something you really want to do and I wanted to get more data about its ability to interfere with progesterone regulation before I started the feeding studies. And so, now that we have reduced the amount of dioxin that it would take to interfere to 10^{-12} molar, I feel like that I should go ahead and do the feeding studies on the animals and see if I get an impact. We have just begun to do that, so we haven't that data studies yet. We only know that in our in vitro model of progesterone suppression, that it takes a very low level of dioxin to interfere with the ability of progesterone to suppress MMPs. Now, as low as 10^{-12} molar, we have been able to see a partial block of progesterone action.

Dr. Suginami: You have shown us the effect of dioxin on the progesterone effect. But there are so many disrupting factors in the environment which we are taking by aspiration or ingestion, and so on and so forth. Could you explain anything about the complex pollution of these agents on human health; reproduction, for instance.

Professor Osteen: Yes, I agree. I guess the day before I left a report came out, and maybe someone read it from the U.S. press about dioxin exposure in certain populations, mostly, I think it was factory workers, had a 60% increase in developing cancer. And I think that came out just this last week. They described that the general population need not worry, because there was no impact on the populations that were not exposed to those high levels. And I think that what people are missing here is that it depends upon the system you are talking about and how vulnerable that system is to an impact. And endometriosis, I think, might be a disease process that might be very, very sensitive to an environmental exposure.

Looking at some of the more recent data, and I think it is in the paper that we are submitting as part of this presentation, that some of the animals that were exposed to dioxin also had differences in their metabolism of other types of agents like the PCBs. And so, you can have this collective type effect down stream that could be caused by one agent affecting the rate of metabolism of another and then maybe perhaps even generational effects like we saw with DES and the next generation. I think that my sense would be that it may be changes that are occurring even during the development that are leading to why the endometrium is different in some women with the disease. Maybe something happened far beyond that that led to that outcome. And, I think that what we don't have is enough specific types of methodology to really look at this disease process. And I think that is what we tried to do with our research. Is there an event that has been linked to the disease that we can directly associate with a level of a toxin that we would be exposed to in our environment? And I think the answer is yes. The question ends up being 'How worried about that should we be?'. And I don't have the answer to that. I think I would like to hear anybody's comment in the room.

I think that affecting the TGF-β expression is such a universal affect in the immune system and the endocrine system that that could be a really a potent factor during development. Because TGF-β is very important in development like processes. If you interfere with its action, that has a lot of potential for downstream problems. But, if anyone in the room has any information or data about dioxin and immune function, especially TGF-β, I think I would love to hear that or hear your thoughts about my data. What do you think that might mean to the disease process?

Professor Terakawa: In the earlier session I understand progesterone induced TGF-β mRNA and the protein expression, and then TGF-β reduced MMP expression, right?

Professor Osteen: That is correct.

Professor Terakawa: In this session, retinoic acid, but not TGF-β, will work to prevent nude mice endometriosis. Can you express once more please the mechanism working for TGF-β?

Professor Osteen: The effect of TGF-β, again, it can be very complex. And what we see in the model is that the model is based on acquiring human endometrial tissue and treating it 24 h in an in vitro system whereby we down-regulate or up-regulate the secretion of MMPs. And, when you take endometrium and treat it with TGF-β, either β_1 or β_2 in vitro, what you see is a biphasic response. You see over the first 24 h a drop in MMP secretion and following the removal of TGF-β over the next 24 h a rebound of MMP secretion occurs that goes higher than the original level. And so, TGF-β seems to have this biphasic effect when the tissue has not also been exposed to progesterone. However, if you treat it with TGF-β and inject it into a mouse, what you are going to see over the next 24 h in the mouse, that the MMPs are going to come back on high and you get just as many lesions as you did with an estrogen treatment. However, that being said, whether we treat endometrial tissue with progesterone or retinoic acid in vitro, the increase in TGF-β expression is involved in the processes by which the MMPs are suppressed. So, there is an interrelationship apparently between the differentiation effect of progesterone and the necessity for TGF-β to act. I think it is complex interaction, and so while TGF-β by itself in the absence of another differentiation factor, either progesterone or retinoic acid, doesn't have a sustained differentiation effect on the tissue, this growth factor is involved in the progesterone environment. So, there is going to be a complex interrelationship. Our interpretation is that you need to have a progestin-like effect in order to get the protection in our model. You cannot get MMP suppression by just TGF-β treatments.

Professor Terakawa: Could you explain the rebound mechanism of TGF-β, the first 24 h less then the gain.

Professor Osteen: You mean the mechanism that explains that effect?

Professor Terakawa: Yes.

Professor Osteen: We have no idea, except for the fact that a similar short-term progesterone exposure leads to a sustained down-regulation of MMPs. So, I think that the TGF-β can have an effect on MMP secretion, but it does not apparently affect the gene expression. And, so, the MMP continues to be translated and more protein is secreted. Again, we don't know exactly at this point how progesterone affects MMP genes. It is the same as the question about how does it affect the TGF-β_2 gene? There are no PREs there. So there may be other factors down stream that are necessary cofactors to get the sustained progesterone effect.

Professor Taketani: Professor Osteen, it looks like you stress the direct action of TCDD to induce the endometriosis. What if the mice were treated and endometrium is implanted in the nude mice given TCDD? Because now the in vitro treatment of TCDD effectively increases the occurrence of endometriosis, if you are actually speaking of the growth of the endometrium implanted in the nude mice. What if the mice were treated and endometrium is implanted in the nude mice which are given TCDD?

Professor Osteen: Yes. We haven't done those studies yet. You mean with the mouse having been exposed?

Professor Taketani: Yes, the mouse is given TCDD, not the tissue.

Professor Osteen: That is the second set of experiments that are being done now. We don't have the results of actually exposing the mice to the toxin and determining if that has an impact on the outcome of the process. Again, we are trying to start those studies now. But, we don't have that data yet.

Professor Taketani: And, another question is how the in vitro treatment of TCDD changes the tissue property to tend towards the development of endometriosis. In other words, how long does the action of the TCDD last? It is a short lived or a long-acting effect?

Professor Osteen: In the explant model, unfortunately, the life of the explant is short, so we can't really ask that question. We will have to ask that question in the nude mice, to see if it has an impact. Our approach is going to be to treat the tissue with the dioxin, establish lesions in nude mice and then attempt to reduce those lesions with either progesterone or retinoic acid treatments to see if you continue to have a long-term resistance to that effect in vivo. Unfortunately these organ cultures, once they are exposed to TCDD, we didn't get any rebound over the life span of the culture, but it is only 4 or 5 days and then the organ cultures begin to break down on their own, so we couldn't study that. We are beginning to look at that same question in the stromal cell culture, and again, we don't know whether or not we are going to be able to get back a progesterone effect downstream. But, we have been working out mainly how low a concentration of TCDD can we get the sustained effect. Certainly lasting up do 6 days or 10 days of culture, but we haven't gone way beyond that to see if we get a sustained affect. And, again, you still wouldn't know if you got the same sort of cell population. So, it is a difficult question to get to.

Professor Taketani: I'd like to ask about the mechanism of the TCDD to induce endometriosis. So far the postulated mechanism by which dioxin induced endometriosis included the impairment of the cellular immunity, the damage to something like that. In your model there is no way to look into the impairment of the cellular immunity because you are using the nude mouse. What do you think about the mechanism related to the immunological derangement?

Professor Osteen: This model by design has positive aspects and negative aspects. The positive aspect of the model is that it allows you to isolate the impact of MMP expression has on the disease process, but you don't have the normal immune components in the model. And so, the model doesn't allow you to actually be able to look at the immune relationships at the same time you are looking at the endocrine relationships. And so I think, again, we need better models to look at immunity. And that is something that I think is very interesting about the aromatase studies that would seem to indicate that the endocrine environment is the dominant factor and that the immune environment might be secondary. I don't know that the aromatase activity has ever been associated with immune function though it might be. I don't know that it has been. But, that data would say that perhaps an estrogen environment or a local estrogen environment is one real key factor in the disease process. Maybe the immune factors are secondary and maybe not as important. I don't know. But, again, I think if you start looking at certain cytokine expressions such as the interleukins and TGF-β, you are looking at cytokines that have roles in both the immune system and in the endocrine system and disrupting their expression or action, I think, will have profound effects on both systems. And, we just need better systems to look at that immune component and I don't think my model will allow us to do that very well.

Dr. Suginami: Professor Osteen, you have mentioned in your earlier lecture that retinoic acid can mimic the progesterone effect and, in this lecture, you have shown through the literature that TCDD decreases retinoic acid induction and increases retinoic acid metabolism. That means TCDD decreases retinoic effect in the tissue. My question is, in your nude mouse model experiment you didn't include a retinoic acid group. Do you have any reason for that?

Professor Osteen: We had focused those studies on the progesterone action by design, but are beginning to do exactly the same thing you say. I think that one of the things that we are setting up right now is the model to look at the isolated stromal cell model in which we can induce retinoic acid synthesis with progesterone. And, as the cells begin to decidualize, you add a retinol precursor and you use HPLC and you can measure increases in retinoic acid. We have set up a series of experiments to determine whether or not TCDD will block that progesterone-induced synthesis of retinoic acid and we don't have those results tabulated yet. But, I think that is a perfect question to ask. And a good question to ask is if the local synthesis of retinoic acid is a component of progesterone action, we want to know whether or not dioxin can block that pathway as well. But, we haven't done that work yet and we are also seeing if we can block retinoic acid action with TCDD the way we have with progesterone. And so, these kinds of studies are in progress, but much of this is new data and we just haven't done everything that needs to be done yet.

Dr. Suginami: I think that retinoic acid is very important because retinoic acid is not only a progesterone mimicker, but also it induces, for instance, apoptosis in some tissues. And, apoptosis can reduce the development of endometriosis, you know. I think that from that point of view, it is, I think, very important.

Professor Osteen: Yes. And we agree and those studies really need to be done.

Professor Hoshiai: Thank you very much for your interesting lecture. I am interested in your experimental design of progesterone and cytokine and MMP models. You didn't mention about the dosage relation in the progesterones and the effect on MMP or on the implantation rate in human endometrium or nude mouse. You have not mentioned about the dose relationship between progesterone and some factors, for example the change of MMPs or change of implantation rate of human endometrium or nude mouse. So, do you have any data that progesterone dose relationship in your models?

Professor Osteen: Just in the model. Obviously we don't have much data in the human model other than our in vitro model. And what you see with progesterone is that the ability to reduce is sort of a complex question. But, in the model, if you have isolated stromal cells and you challenge them with interleukin, you stimulate MMP-3 very high. You can dose out progesterone very easily in terms of its ability to prevent that induction with IL-1. So, it is a really straight dose response curve. And, then you can shift that curve by blocking progesterone with dioxin, so that you take more progesterone to reduce the MMPs. So, you can dose out the retinoic acid and the progesterone and the dioxin so that you can see that there is a dynamic equilibrium that exists. And, to a certain extent, we have just begun to see, could you override dioxin interference by increasing much higher levels of progesterone. Well, you can to a certain extent, but you get to be pretty high levels which are nonphysiologic when you do that. But, certainly what could happen in the endometrium is that if you somehow impair the progesterone's effect then you would dilute the ability to keep the suppression of MMPs. So, you could see why you would get early loss of pregnancy as a confounding issue in women exposed to TCDD. Recurrent pregnancy loss would be something you would predict from this data that you would see. Again, it is very difficult to get the exact data that would do that in our models and we don't have that yet.

Dr. Furuya: Thank you. I have a small question. I am interested about animal model endometriosis experiments. I have never studied nude mice, but I studied endometriosis using rat models. Establishment of autografted endometriotic tissue in the peritoneal cavity in the rat, shows very interesting results of the significantly decreased in the NK cell, natural killer cell, activity in the spleen cells in the rat models. I have never studied the nude mouse model. I think that it is a possibility due to the influence of the intraperitoneal immunologic change including NK cell activity in the nude mice, because it is very important I think in the steroid hormones, as your comment inferred, by steroid hormone and MMP expression, but as a part of the key point of the establishment of the endometriosis, is heterogenic implantation using your models. Could you say if it is possible to influence the NK cell as immunological change?

Professor Osteen: Yes, I think it is. Again, the model offers certain benefits and certain things that are not of benefit. But, absolutely. I think that there are certain changes that we know, certain factors from other studies and other models we know are important and I think what we have to begin to do is add each of those into the model system until we get a full understanding of what all the parameters are. One of the studies that we have going now which is an obvious thing to do is to take ectopic human endometriosis tissue and put it in the nude mouse and try to regulate the growth of that tissue. And so we have those studies going now. Endometriosis is a disease where all these different systems come into play, and I think you have to isolate each one. We are looking at some inhibitors of angiogenesis, for example. Because, if you don't bring in the vasculature, you are not going to get a surviving lesion. So, you might be able to block the model by interfering with neovascularization. In fact, I'm sure you can. Each of these systems is having an impact which needs to be studied independently somehow and I think we just have to design the models a little differently so that we can study each system.

Dr. Kobayashi: You show that dioxin directly or indirectly abrogates progesterone-induced suppression of endometrial development. And you evaluate the sites – you evaluate a number of animals such as were implanted or not. Do you find a difference of size of the lesion that is developed from tissues treated with dioxin?

Professor Osteen: Yes. One of the factors that is a bit of a variable in this; and we are trying to address that question more precisely with some morphometric analysis. The model is based on originally on the premise that human endometrial tissue should be able to, injected in the peritoneum, form these lesions on its own. So, we just inject the tissue. And, so there is an inherent difference in the breakdown of those fragments in the peritoneum just from the model. And so, what we find at 10 days after injection if the tissue is exposed to dioxin, you get the number of lesions per animal is increased and then the average size of the lesion is increased. What we want to do is to develop that into some more of a quantitative-type assessment. So, we are trying to work on what would work with that. Because, you have the numbers of lesions might indicate how many of the injected fragments had the capacity to form an ectopic lesion and then the other question is how aggressive is that lesion in terms of its growth once it has formed and I think we are trying to determine to what degree you can analyze in a semiquantitative way those types of parameters. So, we are looking at that carefully. The best model is reducing lesions with progesterone after so many days of treatment. And I think what we want to do is to say 'Can we get a better quantitative assessment to determine the size of lesions as well as their capacity to develop as a disease?'. So, we have a number of things like this that we want to do with animals with different time points to see how aggressive these lesions will become, but we haven't completed all of those studies yet.

Gynecologic and
Obstetric Investigation

Gynecol Obstet Invest 1999;48(suppl 1):57–58

Summary

Yuji Taketani

We have been fortunate to share in the stimulating and thought-provoking presentations featured at the Takeda-Sponsored 2nd Japan Conference on Endometriosis, held in Kanagawa, Japan on May 8, 1999. There were many significant findings highlighted during the meeting including the role of ovarian hormones and their receptors in the regulation of endometriosis, the regulation of growth factors in the vascular pathogenesis of endometriosis, the role of free radicals in fertility, and new methods of diagnosing endometriosis.

Retinoic acid may act in a manner similar to progesterone in the regulation of endometrial matrix metalloproteinases (MMPs) to reduce the establishment and progression of experimental endometriosis. Professor Kevin Osteen (Vanderbilt University School of Medicine) noted that retinoic acid is known to be an essential component of normal endometrial function, including the maintenance of normal pregnancy. Developing a better understanding of MMP regulation by both steroids and cytokines may help us to develop more successful treatment strategies for endometriosis.

Assistant Professor Jiro Fujimoto (Gifu University) reported that elevated levels of vascular endothelial growth factor (VEGF) secreted into the peritoneal fluid from activated macrophages may contribute to vascular pathogenesis of endometriosis under the influence of ovarian steroids, while basic fibroblast growth factor (bFGF) might promote endometriosis in autocrine and paracrine manners independent of sex steroids. One likely autocrine growth factor for stimulation of the angiogenic pathogenesis of endometriosis endometrial stromal cells is IL-8. Dr Fujimoto believes that many angiogenic factors might contribute individually to the advancement of angiogenic lesions.

Endometriotic implants contain both estrogen receptors and P450arom, an enzyme that catalyzes conversion of androgens to estrogens, suggesting that estrogens may be involved in tissue growth through interaction with the estrogen receptors. Assistant Professor Jo Kitawaki (Kyoto Prefectural University School of Medicine) evaluated the clinical usefulness of a diagnostic test for endometriosis by detection of P450arom in endometrial biopsy specimens and found that P450arom is expressed in the eutopic endometria only in patients with endometriosis. It may soon be possible to diagnose the presence of endometriosis by examination of P450arom expression in endometrial biopsy specimens. This technique may be useful as an initial screening for infertility.

Assistant Professor Hirotaka Ota (Akita University School of Medicine) answered the question 'Do free radicals participate in fecundity?'. He noted that in pregnant mice, the administration of a nitric oxide synthetase inhibitor reduced the pregnancy rate in a dose-dependent manner. Histologic analysis of these uteri showed impaired intercellular reactions and retarded 'decidualization' of stromal cells. We can conclude that free radicals appear to participate intimately in implantation, and the optimal levels of free radicals are critical to the process.

KARGER

Fax + 41 61 306 12 34
E-Mail karger@karger.ch
www.karger.com

Accessible online at:
www.karger.com/journals/goi

The pathogenesis and activity of endometriosis may depend on internal angiogenesis, and may be different in different classifications of pigmented lesions in pelvic endometriosis. Dr. Akira Fujishita (Nagasaki University) reports that high endoglin levels may provoke an increase in the area of the subperitoneal blood vessel network, which may in turn facilitate implantation and viability in the retroperitoneal space. In contrast, lower endoglin levels in black and white lesions could explain the decrease in stromal vascularization and subsequent inactivation of the implant. He noted that immunohistochemical studies of endoglin expression may be better than VEGF expression for understanding the activity and pathogenesis in endometriosis.

Professor Osteen also shared his findings on the possible association between endometriosis and environmental exposure to dioxin (TCDD). TCDD exposure inhibits the expression of progesterone-associated transforming growth factor (TGF)-β_2 and suppresses associated matrix metalloproteins. Interference with the production of TGF-β_2 may have broad implications to reproductive processes beyond development of endometriosis. It is obvious that a better understanding of progesterone and the role of TCDD is needed.

This symposium included some of the top expert investigators from the field of endometriosis. It is hoped that the studies featured in this supplement will be the basis for a new understanding of cellular and molecular function in the development of endometriosis. These new findings can advance our understanding of this disease and aid in our development of new therapies for the prevention and successful treatment of endometriosis.